天津国际设计周建筑展
(2017—2019)
Architecture Exhibition of Tianjin International Design Week
(2017-2019)

宋　昆　　李云飞　　胡子楠　　著
Song Kun　　Li Yunfei　　Hu Zinan

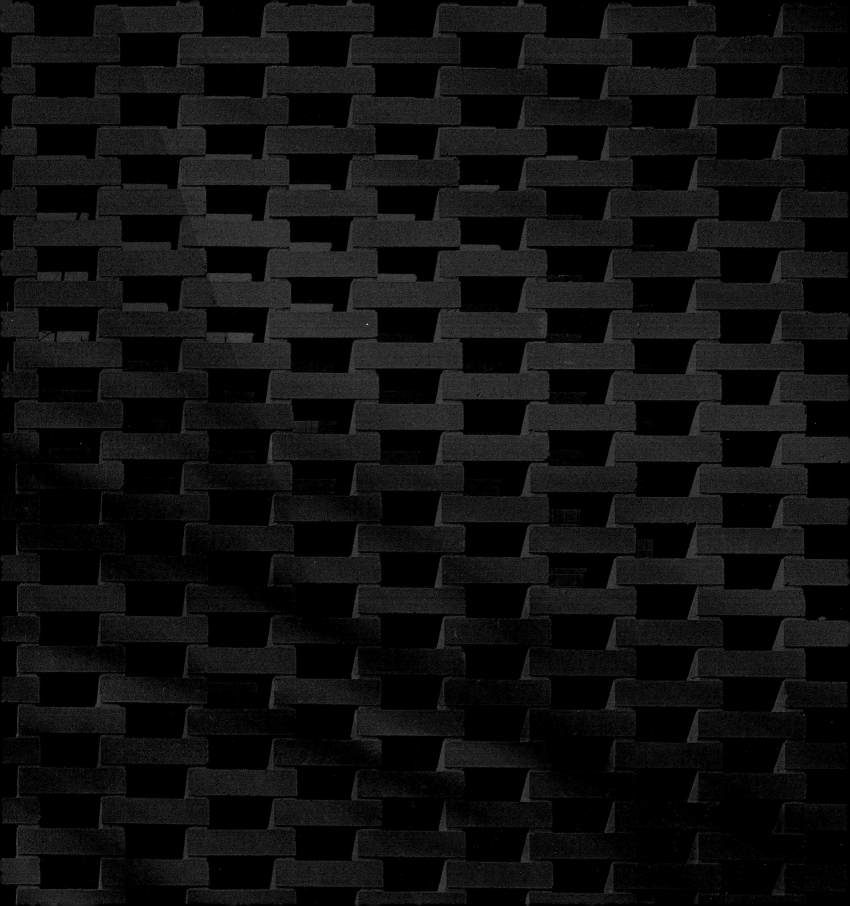

目录
Content

天津国际设计周建筑展（2017—2019）
Architecture Exhibition of Tianjin International Design Week (2017-2019)

导言 Introduction	1
黑川雅之 Masayuki Kurokawa	2
宋昆 Song Kun	4
马西莫·马雷利 Massimo Marrelli	6
李云飞 Li Yunfei	8

2017年建筑展（装置展） 10
2017 Architecture Exhibition (Installation Exhibition)

当代中的非当代——2017天津国际设计周建筑展综述 14
Non-contemporaneity in Contemporaneity—Summary of Architecture Exhibition of 2017 Tianjin International Design Week

对天津国际设计周执行主席李云飞的访谈 19
An Interview with Li Yunfei, the Executive Chairman of Tianjin International Design Week

2017参展建筑师及作品 25
2017 Participant Architects & Their Works

陈天泽 甄明扬（天津市建筑设计研究院有限公司） 27
Chen Tianze, Zhen Minyang (Tianjin Architecture Design Institute Co., Ltd.)

顾志宏（天津大学建筑设计规划研究总院有限公司/顾志宏工作室） 31
Gu Zhihong (Tianjin University Research Institute of Architectural Design and Urban Planning Co., Ltd./Gu Zhihong Studio)

那日斯（天津市博风建筑工程设计有限公司） 35
Na Risi (Tianjin BF Architecture Design Co., Ltd.)

任军（天津市天友建筑设计股份有限公司） 39
Ren Jun (Tianjin Tenio Architecture and Engineering Co., Ltd.)

宋昆 胡子楠（天津大学建筑学院） 43
Song Kun, Hu Zinan (School of Architecture, Tianjin University)

王振飞（HHDFUN事务所） ... 47
Wang Zhenfei (HHDFUN Architects)

徐强（天津天华建筑设计有限公司） ... 51
Xu Qiang (Tianjin Tianhua Architecture Planning & Engineering Ltd.)

赵劲松（天津大学建筑学院/非标准建筑工作室） ... 55
Zhao Jinsong (School of Architecture, Tianjin University/Non-standard Architecture Studio)

张大昕（天津大学建筑设计规划研究总院有限公司） ... 59
Zhang Daxin (Tianjin University Research Institute of Architectural Design and Urban Planning Co., Ltd.)

张曙辉（北京八作建筑设计事务所有限公司） ... 63
Zhang Shuhui (Beijing Bazuo Architecture Office, Ltd.)

2018年建筑展（作品展） ... 66
2018 Architecture Exhibition (Projects Exhibition)

此间——2018天津国际设计周建筑展综述 ... 70
In Between—Summary of Architecture Exhibition of 2018 Tianjin International Design Week

2018参展建筑师及作品 ... 73
2018 Participant Architects & Their Works

卞洪滨（天津大学建筑学院） ... 75
Bian Hongbin (School of Architecture, Tianjin University)

王宽（宽建筑工作室） ... 79
Wang Kuan (KUAN Architects)

任军（天津市天友建筑设计股份有限公司） ... 83
Ren Jun (Tianjin Tenio Architecture and Engineering Co., Ltd.)

庄子玉（BUZZ庄子玉工作室） ... 87
Zhuang Ziyu (BUZZ/Büro Ziyu Zhuang)

张华（天津大学建筑设计规划研究总院有限公司/张华工作室） ... 93
Zhang Hua (Tianjin University Research Institute of
Architectural Design and Urban Planning Co., Ltd./Zhang Hua Studio)

张曙辉 王淼（北京八作建筑设计事务所有限公司） ... 97
Zhang Shuhui, Wang Miao (Beijing Bazuo Architecture Office, Ltd.)

卓强（天津市建筑设计研究院有限公司） ... 101
Zhuo Qiang (Tianjin Architecture Design Institute Co., Ltd.)

赵劲松（天津大学建筑学院/非标准建筑工作室） ... 105
Zhao Jinsong (School of Architecture , Tianjin University/Non-standard Architecture Studio)

鲍威（BWAO/鲍威建筑工作室） ... 109
Bao Wei (BWAO/Bao Wei Architecture Office)

2019年建筑展（作品展） 112
2019 Architecture Exhibition (Projects Exhibition)

汇流而下的城乡之辩——2019天津国际设计周建筑展综述 116
The Confluence: the Debate between Urbanity and Rurality—Summary of Architecture Exhibition of 2019 Tianjin International Design Week

2019参展建筑师及作品 121
2019 Participant Architects & Their Works

张华（天津大学建筑设计规划研究总院有限公司/张华工作室） 123
Zhang Hua (Tianjin University Research Institute of Architectural Design and Urban Planning Co., Ltd./Zhang Hua Studio)

赵劲松（天津大学建筑学院/非标准建筑工作室） 127
Zhao Jinsong (School of Architecture, Tianjin University/Non-standard Architecture Studio)

狄韶华（第一实践建筑设计） 131
Di Shaohua (PRAXiS d'ARCHITECTURE)

任军（天津市天友建筑设计股份有限公司） 135
Ren Jun (Tianjin Tenio Architecture and Engineering Co., Ltd.)

郭海鞍（中国建筑设计研究院乡土创作研究中心） 139
Guo Hai'an (Rural Culture D-R-C, China Architecture Design & Research Group)

韩文强（中央美术学院建筑学院/建筑营设计工作室） 143
Han Wenqiang (School of Architecture, Central Academy of Fine Arts/ARCH STUDIO)

陈天泽 关英健 甄明扬（天津市建筑设计研究院有限公司） 147
Chen Tianze, Guan Yingjian, Zhen Mingyang (Tianjin Architecture Design Institute Co., Ltd.)

卜骁骏 张继元（时境建筑设计事务所） 151
Bu Xiaojun, Zhang Jiyuan (Atelier Alter Architects PLLC)

田恬（天津市城市规划设计研究总院有限公司） 155
Tian Tian (Tianjin Urban Planning & Design Institute Co., Ltd.)

张曙辉 王淼（北京八作建筑设计事务所有限公司） 159
Zhang Shuhui, Wang Miao (Beijing Bazuo Architecture Office, Ltd.)

王求安（北京安哲建筑设计有限公司） 163
Wang Qiu'an (Beijing ANT Architectural Design Co., Ltd.)

张东光 刘文娟（合木建筑工作室） 167
Zhang Dongguang, Liu Wenjuan (Atelier Heimat)

申江海（大观建筑设计） 171
Shen Jianghai (Daga Architects)

后记 174
Postscript

导言
Introduction

黑川雅之
天津国际设计周总顾问

Masayuki Kurokawa
General Consultant of Tianjin International Design Week

献给传统与未来的赞歌
A Hymn to Tradition and Future

文化难于定义。我以为文化是"记忆和愿望的叠加"。人们总是无法从记忆中逃离,但又总是对未来抱有美好的愿望。在某一个瞬间,人们会被过去的记忆牵绊,与此同时又会有一股无论如何也要奋勇向前的强烈意愿支配着内心。这就是救赎。

在美丽而节奏快速的城市背后,有着让人怀念和感到温暖的历史;在拥有美好内心的人背后,是那一抹过去的影子,赋予其更加迷人的魅力。当然,如果城市和人不能同时具备挑战未来的姿态,那就不能认为他们具有真正的魅力。

天津大学是一所有着优良传统的大学。学生中人才辈出,教师队伍同样人才济济,想必大家都是被这所大学的深厚的历史底蕴吸引而来的吧。天津大学同天津这座城市一样,拥有着丰富的历史和美好的未来。

魅力的形成并非一蹴而就,在魅力的背后是过往积蓄的力量。天津大学的建筑师展会与在北宁公园举办的天津国际设计周的内容相契合,已经连续举行3年;天津大学是天津市的代表,既拥有悠久历史,又面向未来,展出的优秀作品也为天津国际设计周创造了新的历史。

建筑是文化的典型成果。建筑造就了街道和城市,让人们的生活富有韵味,层层叠加创造着城市的文化。面向未来的行动正是在书写未来的记忆。

我相信,天津大学和天津国际设计周的活动,不仅创造了天津过往的历史,而且叠加了对未来的伟大愿景。

由衷祝贺本书的出版。

Culture is difficult to define. In my opinion, culture is "the accumulation of memory and desire". People always fail to escape from their memories, but they still yearn for the future. At some moment, people are bound by their memories of the past, but at the same time, they still have a strong desire to move forward. This is redemption.

Behind the beautiful and fast-paced city, there is a history that makes people feel nostalgic and warm; behind a person with a heart of gold, it is the shadow of the past that makes him more attractive. Of course, if the city and people cannot challenge the future simultaneously, we cannot say they have authentic charm.

Tianjin University has a fine tradition. There are not only a large number of talented students but also many talented teachers. The deep history background of this university attracts many people. Like Tianjin, Tianjin University has a rich history and a bright future.

Charm cannot be achieved overnight because behind it is the power accumulated in the past. The architects' exhibition of Tianjin University is very suitable for Tianjin International Design Week held in Beining Park, and it has been held in Tianjin Design Week for three consecutive years. As a representative of Tianjin City, Tianjin University has a long history and faces the future. The outstanding works on display have created a new history for Tianjin International Design Week.

Architecture is a typical achievement of culture. Architecture creates streets and cities. Owing to architecture, people's life is full of charm, and urban culture is created layer by layer. Future-oriented action is writing the memory of future.

I believe Tianjin University and Tianjin International Design Week will create the history of Tianjin and accumulate great wishes for the future.

Congratulations on the publication of this book.

宋 昆

天津大学建筑学院党委书记、院长、教授、博士生导师
天津国际设计周建筑展总策展人

Song Kun

Secretary of the Party Committee, Dean, Professor, Doctoral Supervisor of the School of Architecture, Tianjin University
Chief Curator of Architecture Exhibition for Tianjin International Design Week

天津国际设计周始于2014年。2017年5月，天津国际设计周建筑展作为以建筑师为创作主体的独立主题展首次出现在天津国际设计周，之后成为设计周最为活跃的版块之一，展览时间为每年的5月份，地点在天津北宁公园内。2020年，因新冠肺炎疫情席卷全球的客观因素其中断举办。

2017年，天津国际设计周的主题为"未来就是现在"，旨在通过设计师对未来与现在的思辨发现设计力量的源泉。为了响应这样的主题，策展团队将本次建筑展的主题命名为"当代中的非当代"，展览由12位建筑师的10件装置作品组成，建筑师包括顾志宏、徐强、赵劲松、王振飞、张大昕、张曙辉、那日斯、陈天泽、甄明扬、宋昆、胡子楠、任军。

2018年，天津国际设计周的主题为"东方×西方"，希望在新时代背景下展示东方与西方设计文化之间的对话与交融。此年度天津国际设计周建筑展的主题确定为"此间"，展览内容为建筑师实践作品，邀请了京津冀地区的9组（10位）建筑师来展示他们近期的建筑思考与实践，他们分别是卞洪斌、王宽、任军、庄子玉、张华、张曙辉/王淼、卓强、赵劲松、鲍威。

2019年，天津国际设计周的主题为"创意：源于城市，发展城市"，旨在探讨创意与城市之间的互动关系，以及创意之于城市的重要作用。本年度天津国际设计周建筑展聚焦天津城市文化及其对设计的影响，选定的主题为"汇流"，张华、郭海鞍、陈天泽/关英健/甄明扬、田恬、任军、狄韶华、卜骁骏/张继元、张曙辉/王淼、王求安、张东光/刘文娟、申江海、赵劲松、韩文强13组（18位）建筑师构成了该年度参展建筑师的阵容。此外，该展览于同年9月参加了北京国际设计周设计之旅版块，赴北京中华世纪坛继续展出。

天津国际设计周建筑展的策展团队以天津大学建筑学院为主体，我为总策划及主策展人，策展团队主要成员包括张昕楠、胡子楠、赵伟等。2019年，ADA（建筑设计艺术）研究中心黄元炤加入。

天津国际设计周建筑展历年主题鲜明，积极探索当下我国城乡发展过程中的热点问题。纵观3年的建筑展，内容包含1年的装置展与2年的实践作品展，展览规模逐年扩大，参展人数逐年增加。参展建筑师以京津冀地区崭露头角的中青年建筑师为主体，他们来自设计院、独立事务所、高校等机构。未来，天津国际设计周建筑展仍将聚焦不同机构里中青年建筑师的设计实践，并希望通过长期积累，以不同主题与形式呈现在公众面前，逐步成为塑造地区建筑文化的重要力量。

Tianjin International Design Week (TDW) has been successfully held for the first time in 2014. In May 2017, the Architecture Exhibition of TDW was first launched as an independent themed exhibition focusing on architects, which has grown into one of the most vibrant TDW sections. It is held every May in Beining Park, Tianjin. Because of the outbreak of COVID-19, it was interrupted from 2020.

2017 TDW themed with "The Future Is Now" focused on the source of design power as designers contemplate the future and the present. In response, curators named the Architecture Exhibition of 2017 TDW with "Non-contemporaneity in Contemporaneity". The exhibition showcased 10 pieces of devices by 12 architects, including Gu Zhihong, Xu Qiang, Zhao Jinsong, Wang Zhenfei, Zhang Daxin, Zhang Shuhui, Na Risi, Chen Tianze, Zhen Mingyang, Song Kun, Hu Zinan, and Ren Jun.

With the theme of "East×West", 2018 TDW showed the connection and fusion of Eastern and Western design culture in a new era. The Architecture Exhibition of this year was themed with "In Between". It displayed real-life works and ideas from 10 architects in the Beijing-Tianjin-Hebei Region, namely Bian Hongbin, Wang Kuan, Ren Jun, Zhuang Ziyu, Zhang Hua, Zhang Shuhui/Wang Miao (design team), Zhuo Qiang, Zhao Jinsong and Bao Wei.

Themed with "Creativity for City, City of Creativity", 2019 TDW shed light on the interplay between creativity and the city, and the significance of creativity to the city. Under the theme of "Confluence", the Architecture Exhibition presented Tianjin's urban culture and its effect on design. Eighteen contributing architects included Zhang Hua, Guo Hai'an, Chen Tianze/Guan Yingjian/Zhen Mingyang (design team), Tian Tian, Ren Jun, Di Shaohua, Bu Xiaojun/Zhang Jiyuan (design team), Zhang Shuhui/Wang Miao (design team), Wang Qiu'an, Zhang Dongguang/Liu Wenjuan(design team), Shen Jianghai, Zhao Jinsong, and Han Wenqiang. In addition, the display continued in September in the China Millennium Monument, as part of Design Hop, a section of 2019 Beijing International Design Week.

The curators of the Architecture Exhibition of TDW mostly come from the School of Architecture, Tianjin University. I am the chief organizer and curator. Others include Zhang Xinnan, Hu Zinan, Zhao Wei, etc. Huang Yuanzhao from ADA Research Center, joined the team in 2019.

The architecture exhibitions of TDW have probed into hot topics that emerge from China's urban and rural development with distinctive themes. With one year for devices and two for real-life cases, the three architecture exhibitions grew in size and participant number. Contributing architects are primarily young and middle-aged architects emerging in the Beijing-Tianjin-Hebei Region, who work at design institutes, firms, and universities. In the future, it will continue to focus on design practice of young and middle-aged architects from different organizations and sectors, and appear in public with new themes and forms. Guided by experience over the years, the event, I hope, will gradually become an influential force in local architectural culture.

马西莫·马雷利

那不勒斯费德里克二世大学前任校长、
公共经济学教授

Massimo Marrelli

Professor of Public Economics
Former Rector
University of Naples Fedrico II

本书收集了2017—2019年在天津国际设计周展示的建筑作品。多年来，天津国际设计周见证了天津大学和那不勒斯费德里克二世大学之间的合作，而我担任那不勒斯费德里克二世大学的校长已5年有余。这对我来说非常重要，会让我想起我们之间合作、创作的美好时刻。作为一名经济学家，每当阅读和浏览这本书的时候，我总会思考这些问题：建筑到底是什么？建筑真正需要什么？建筑的本质又是什么？

这本书给人的第一印象便是：它呈现给人们的是"好的建筑"，而非"坏的建筑"。有人不禁会问："为什么？"建筑的文学定义是设计的艺术和科学，以及具有美学特征的建筑空间、结构和环境，它们存在的目的是带给观众感官上的刺激。这个定义从多学科的角度揭示了一些基本事实。首先，建筑是一门艺术。因为建筑设计师多是具有建筑天赋和才能的艺术家，他们通过作品表达自己的创造力。其次，建筑是一门关于设计、建造空间、结构或建筑的科学，涉及多门学科的结合，包含数学、物理和其他相关的科学和科学方法，其中还涉及一些别的领域。因此，当我看到这本书收录的作品时，便提到了"好的建筑"。这本书收集了倾向于形成和建立"共享文化"的想法和项目。

"共享文化"是指位于时间和空间中的物理形态和虚拟形态的文化，这种文化在具有社会凝聚力的社区得以共享和表达。共享文化是在特定的地理或虚拟区域内可用的智力资源系统，我们可以认为它是更传统意义上的文化区或文化集群概念的演变。一个社区的创意、创造力和风格，传统领域知识，信仰，仪式和习俗，共享和参与式生产技术等定义了共享文化。例如：城市形象、当地语言、巴罗洛葡萄酒的品牌、艺术运动、人们在网络上生成的内容、本土社区掌握的传统知识以及设计师和艺术家社区表达的创造力。

公共空间和文化共享空间在理论上有哪些主要区别？作为公共产品，文化共享空间的承载能力是无限的：消费文化并不会减少其他文化的总量，在消费方面不会构成竞争。一首音乐或一首诗，人们可以不受限制地消费、欣赏和聆听，景观或都市环境也是如此。公共文化的资源是不会枯竭的。

众所周知，公共自然资源的承载能力却是有限的 (Hardin,1968; Ostrom, 1990, 1992, 2002)。个人因私人利益获取公共资源而不顾及对公共利益的负面影响，这就需要政府来规范经济活动、私有财产或自组织的资源 (Ostrom, 1990; Lam, 1998)。相反，文化共享空间不会因此受损。

当代城市在后工业化市场的先进生产和服务中扮演着新的角色。城市在吸引自由投资、使公司和企业本地化、申办国际赛事、吸引旅游流或留住人口和人力资本方面展开竞争。财富的创造越来越以创新、创造力、文化等非物质因素为基础。但是，只有当城市成为具有文化共享特征的共同资源，即在进化过程中，作为当地社区共享的文化和价值观的结果时，才能取得成功。

这本书提出的观点与文化共享的观点一致。不同的展览标题可能引发对这一解读的探讨：

2017年主题——"当代中的非当代"；

2018年主题——"此间"；

2019年主题——"汇流"。

共享进化过程开始于公共文化基础。

对于天津大学贡献的重要展览和精美书籍，我在此表示衷心的感谢！

This book collects architectural works presented in 2017-2019 TDW. TDW has witnessed, over the years, a collaboration between the University of Tianjin and the University of Naples Federico II of which I have been Rector for 5 years. For me, it is particularly important and brings to mind wonderful moments of collaboration and creation. As an economist, when reading and browsing this book, I always think about these questions：what on earth is architecture?What does architecture really need? What is the essence of architecture?

The first impression that the book gives people is that what is presented is "good architecture" as opposed to "bad architecture". Someone may ask: "why?" In the literature, architecture is defined as the art and science of designing as well as architectural space, structure and surroundings with aesthetic features. The purpose of its existence is to give viewers a sense of excitement. This definition brings out a few fundamental facts from a multidisciplinary perspective. Firstly, it is an art because most architects are artists with talent and aptitude for architecture, who express the creative ability through their works. Secondly, it is a science of designing and building space, structure or building. This process involves a combination of multiple disciplines, such as mathematics, physics and other related sciences and scientific methodologies. Meanwhile there is something else. That is why I speak of "good architecture" when I see the works in this book. This book collects ideas and projects that tend to form and build "Cultural sharing".

"Cultural sharing" refer to the culture located in time and space —either physical or virtual — and shared and expressed by a socially cohesive community. Cultural sharing is a system of intellectual resources available in a given geographical or virtual area. It can be thought of as the evolution of the more traditional concept of cultural district or cultural cluster. Ideas, creativity and styles of a community, traditional knowledge, beliefs, rites and customs, shared and participated productive techniques define cultural sharing. Some examples are: the image of a city, a local language, the brand of Barolo wine, an artistic movement, the content generated by users on the Web, traditional knowledge held by indigenous communities, and the creativity expressed by designers' and artists' communities.

Which are the main theoretical differences between commons and cultural sharing? The carrying capacity of cultural sharing is endless: consuming culture does not reduce its total amount for the others. They are non-rival in consumption. A music or a poem can be consumed, appreciated and listened to without any limit, so is landscape or urban environment. There is no exhaustion of the common cultural resources.

As is known a natural common resource has a limited carrying capacity (Hardin, 1968; Ostrom, 1990,1992,2002). Individuals with private interests and open access do not take into account their negative effects on the common interests. This calls for a government to regulate the economic activities, private property or self-organized resource(Ostrom, 1990; Lam, 1998). Cultural sharing on the contrary doesn't suffer from this problem.

Contemporary cities have new roles in advanced production and services in post-industrial markets. Cities compete in attracting footloose investments, in localizing firms and corporations, in bidding for hosting international events, in attracting touristic flows or in retaining population and human capital. The creation of wealth is more and more based on immaterial factors, such as innovation, creativity and culture. But they can succed if and only if cities become a common resource with the characteristics of cultural sharing, i.e. as the outcome of common culture and values shared by local communities in the evolutionary process.

The ideas presented in this book are coherent with the idea of cultural sharing. The headings of different exhibitions evoke this possible reading/interpretation:

"Non-Contemporneity in Contemporaneity"— the theme in 2017；

"In Between" — the theme in 2018；

"Confluence" — the theme in 2019。

Shared evolutionary process starts from common cultural foundations.

I would like to thank Tianjin University for these very important exhibitions and the beautiful book.

李云飞
天津国际设计周执行主席

Li Yunfei
Executive Chairman of Tianjin International Design Week

天津国际设计周(TDW)与天津大学(以下简称"天大")建筑学院的渊源可追溯到2014年首届设计周活动。作为初创团队中重要的一员，天大建筑学院与天津国际设计周组委会已携手走过九个年头。在此过程中，天大建筑学院始终是这个团队的中坚力量。作为中国建筑名校，天大在中国建筑学术领域有很高的水平。从设计周的展览到国际班再到论坛等各类会展活动，正因为有了天大师生的深度参与，才使天津国际设计周有信心也有底气将中国的设计教育成果展现在世界面前。

从2017年的第四届天津国际设计周开始，天大建筑学院连续3年举办了建筑设计主题展览。从"当代中的非当代"到"此间"再到"汇流"，天大建筑学院的每一次展览都带来了不一样的惊喜。由于展览组织方有意识地组织不同性质的设计团体参展，因此呈现出多样化背景下设计师对同一设计主体的不同诠释，使我们有机会全面领略中国建筑设计的水平与它背后的精神。在设计主题的把握上，建筑展既完美贴合并延伸了设计周的总主题，又展现了其对中国建筑设计的深入思考与探索。而在展览的内容与形式上，则充分体现了建筑思想的多样化，从关注当下到思考未来，从城市建设到乡村振兴，从现实案例到概念装置，从不同角度呈现了建筑坚硬外壳之下那有趣的灵魂。可以说，天大的每一次主题展都是设计周最值得期待的展览之一，更是最能够代表中国建筑设计的展览之一。

很荣幸与天大建筑学院这样优秀的团队携手走过7年，在这7年里怀着为天津、为中国设计发声的共同理想，我们彼此扶持，走得更远、更好。

欣闻天大建筑学院将这3年的主题展览结集出版，我由衷地感到高兴与自豪，将这些合作的结晶以这样的形式记录下来并让更多人看到，也是天津国际设计周组委会一直以来的愿望。在此我谨代表天津国际设计周团队，对此书的出版致以衷心的祝贺，也希望未来我们能够共同挖掘中国更多优秀的建筑设计作品，让中国设计被世界看到。

The ties of Tianjin International Design Week (hereinafter called TDW) and the School of Architecture, Tianjin University (hereinafter called TJU), are dated to the first TDW in 2014. As a vital member and backbone of the startup team, the School has worked closely with the TDW organizing committee for nine years. As a famous university of architecture, TJU is at a very high level in the academic field of Chinese architecture. Thanks to the profound contributions made by TJU's faculties and students to TDW, such as exhibitions, workshops and forums, and we thus are confident to show the world China's progress in design education.

Since the 4th TDW in 2017, the School of Architecture, TJU has hosted architecture exhibitions for three years. From "Non-contemporaneity in Contemporaneity" in 2017 to "In Between" in 2018, and then to "Confluence" in 2019, each exhibition of the School of Architecture, TJU brought a different surprise to the audience. Since the TDW organizers intended to bring in design groups of different natures, there are various expressions over the same subject by designers from diverse backgrounds, which gives us a chance to understand how far the Chinese architectural design has gone and the spirit behind it. As for the themes, the Architecture Exhibition perfectly matched and extends TDW's general theme and reflects its deep thought and exploration of Chinese architecture and design; its works embody diverse architectural concepts, focusing on the present, the future, urban development, and rural revival. Through real-life cases and concept devices, It demonstrates the essence of architecture from different perspectives. It is safe to say that each themed exhibition by TJU is among the much-awaited shows during TDW and representative of Chinese architectural design.

It has been a great honor to cooperate with such a remarkable team as the School of Architecture, TJU for seven years, during which time, we have supported each other for more progress in pursuing our shared dream of advocating for Tianjin and Chinese design.

I am pleased and proud to know that the last three years' works of theme exhibitions would be compiled and published by the School of Architecture, TJU. These exhibitions are born out of our partnership, and to record them in a book and make more people see them has always been the dream of organizing committee. On behalf of the TDW organizers, I wish this book a great success and hope our cooperation will lead us to more extraordinary architectural designs and raise the profile of the Chinese design at a broader stage.

2017 年建筑展（装置展）
2017 Architecture Exhibition (Installation Exhibition)

2017

2017年建筑展
（装置展）
2017 Architecture Exhibition
(Installation Exhibition)

当代中的非当代
——2017天津国际设计周建筑展综述
Non-contemporaneity in Contemporaneity
—Summary of Architecture Exhibition of 2017
Tianjin International Design Week

对天津国际设计周执行主席李云飞的访谈
An Interview with Li Yunfei,
the Executive Chairman of Tianjin International Design Week

2017参展建筑师及作品
Participant Architects & Their Works

陈天泽 甄明扬（天津市建筑设计研究院有限公司）
Chen Tianze, Zhen Minyang (Tianjin Architecture Design Institute Co., Ltd.)

顾志宏（天津大学建筑设计规划研究总院有限公司/顾志宏工作室）
Gu Zhihong (Tianjin University Research Institute of Architectural Design and Urban Planning Co., Ltd./Gu Zhihong Studio)

那日斯（天津市博风建筑工程设计有限公司）
Na Risi (Tianjin BF Architecture Design Co., Ltd.)

任军（天津市天友建筑设计股份有限公司）
Ren Jun (Tianjin Tenio Architecture and Engineering Co., Ltd.)

宋昆 胡子楠（天津大学建筑学院）
Song Kun, Hu Zinan (School of Architecture, Tianjin University)

王振飞（HHDFUN事务所）
Wang Zhenfei (HHDFUN Architects)

徐强（天津天华建筑设计有限公司）
Xu Qiang (Tianjin Tianhua Architecture Planning & Engineering Ltd.)

赵劲松（天津大学建筑学院/非标准建筑工作室）
Zhao Jinsong (School of Architecture, Tianjin University/Non-standard Architecture Studio)

张大昕（天津大学建筑设计规划研究总院有限公司）
Zhang Daxin (Tianjin University Research Institute of Architectural Design and Urban Planning Co., Ltd.)

张曙辉（北京八作建筑设计事务所有限公司）
Zhang Shuhui (Beijing Bazuo Architecture Office, Ltd.)

当代中的非当代——2017天津国际设计周建筑展综述
Non-contemporaneity in Contemporaneity —Summary of Architecture Exhibition of 2017 Tianjin International Design Week

总策展人：宋昆
Chief Curator: Song Kun

策展人：赵劲松、张昕楠、胡子楠、赵伟、苑思楠
Curators: Zhao Jinsong / Zhang Xinnan / Hu Zinan / Zhao Wei / Yuan Sinan

2017年5月，天津国际设计周建筑展在天津北宁公园开幕，这是天津国际设计周举办四届以来，首次以建筑师为创作主体而设立的独立主题展。策展团队来自天津大学建筑学院，团队结合当代中国的发展现状，试图通过"当代中的非当代"这一命题来探讨快速城市化背景下传统与未来、地域性与全球化、生态环境、"互联网+"等问题为建筑学学科所带来的机遇与挑战。本次展览邀请了京津两地的中青年建筑师参加，作品以构筑物、装置、虚拟现实等形式呈现出他们的独特立场与长期思考。

一、展览缘起与策展过程

天津国际设计周是我国北方地区规模较大的创意设计类展览，汇聚了中国、意大利、日本、德国、荷兰等国的优质设计资源，在推动本地区创意设计产业发展以及提高大众审美意识等方面做出了突出的贡献。自2014年起，天津国际设计周已成功举办九届，旨在探讨设计活动与日常行为、时代变革等热点话题的联系，主题包括"随艺生活""记忆与梦想""行住坐卧""未来就是现在"等。天津国际设计周将目标群体指向决策者、国内外设计师、学生以及普通民众，逐步形成了以展览、竞赛、论坛、工作坊、研讨会等为载体的多元模式，构建起天津地区城市文化交流的重要平台。

建筑师是对时代、社会、城市、文化具有敏锐洞察力的重要群体，以他们为创作主体的独立展览于2017年首次加入天津国际设计周。该群体希望透过建筑学的阅读视角，采用设计策略回应本次设计周的主题。

本次天津国际设计周建筑展的总策展人为天津大学建筑学院宋昆教授，团队成员包括赵劲松、张昕楠、胡子楠、赵伟、苑思楠。策展团队于2016年10月开始筹备，初期工作包括拟定主题、展览形式及受邀建筑师名单；2016年11月，确定了与2017年天津国际设计周主题"未来就是现在"相对应的建筑展主题——"当代中的非当代"，并邀请了12位建筑师参展，他们出生于20世纪60年代末到80年代初，任职单位涵盖国有建筑设计院、民营企业、独立事务所、高校等，能够体现新一代建筑师在理论和实践方面的敏锐度和洞察力；2016年12月，参展建筑师前往天津北宁公园勘察现场，选定展位；2017年2月，参展建筑师在天津大学建筑学院进行了第一次方案研讨会；2017年3月，天津国际设计周总顾问、日本工业与建筑设计大师黑川雅之先生会见参展建筑师，并就展览内容进行了沟通；2017年4月上旬，参展建筑师在天津巷肆创意产业园进行了第二次方案研讨会，基本确定了参展作品方案与施工计划；2017年4月下旬，建筑师开始进场建造各组参展作品，宣传工作也随之开始；2017年5月12—17日，展览正式举行。

二、从"未来就是现在"到"当代中的非当代"

2017天津国际设计周以"未来就是现在"为主题，意图在现在与未来的衔接转化与辩证思考中汲取新的设计力量。"现在"是漫步经过的一条道路，是眼前掠过的一个映象，是手心触碰的一丝冰冷，是身上汇聚的一束光线，它切实、客观，充斥在我们的四周，以"物"的方式证明自身的存在。"未来"是迷茫中的一种寻找，是混沌中的一刻反思，是喧闹中的一丝静默，是黑暗中的一片梦境，它鲜活、主观，孕育于我们的脑海中，通过"思"来呈现独特的力量。

面对这样的主题，策展人认为"设计"是"物"与"思"的融合，一方面是形而下对于现实、物性的把握，另一方面则是形而上对于世界、自我的认知。"现在"是对已知的肯定，是身体在场的知觉，是此时此地的存在于世，它扑面而来，是一种"为我"的存在。"未来"是对未知的好奇，是意象呈现的期许，是由此及彼的另一天地，它萦回环绕，是一种"超我"的体验。"设计"是"为我"与"超我"的融合，容纳和唤醒人们的感知与想象。"现在"与"未来"是对时间的度量。每一个现在都是曾经的未来，每一个未来都是下一个现在。所谓设计，总是同时存在于此刻的情状和下一刻的预示中，是"生活时间"的映射，是"绵延"。人们面对现在，思考未来；面对未来，思考现在。

基于以上，策展团队将本次建筑展的主题命名为"当代中的非当代"。在建筑学语境下，当代性一直以来都是值得深思和探讨的话题，"当代"可以理解为一种对"现在"的反思和批评，"非当代"在某种意义上则包含了成就"现在"的传统记忆，以及影射"未来"的明日愿景。策展团队希望将这份略带哲学意味的思辨与当下我国城市发展的现状相结合，并通过建筑师的作品来寻找不同的答案。

三、展览概况

本次建筑展由12位建筑师的10件作品组成，场地位于天津北宁公园中的两座二层中国传统形制的建筑及其之间的场地区域之内。场地类型包括传统建筑的室内房间、室外庭院以及两座建筑之间的水面空间，参展建筑师分别以传统、未来、空间、时间、边界等不同视角为切入点进行创作，从不同维度呈现对于主题的思考与诠释。

作为对基地的回应，"传统与当代"自然成为对设计命题的重要解读之一。顾志宏的作品《红亭子》基于庭院中一处中国传统形制的凉亭进行设计，将亭子的形式逐渐裂化，通过亭子上下的鲜明对比，在传统的单纯宁静与当代的多变纷扰间传递出关于人们对生存状态的种种困惑与反思；徐强的作品《时间的柱廊》将西方古典空间元素引入中国传统院落，以当代材料的特有属性对传统材料进行反向映射，从而有意营造一种抽象与具象的冲突和并置，激发传统与当代的对话。

"当代中的非当代"除涉及传统外，还面向未来，本次建筑展关于"当代与未来"的思考更多是源于建筑师对人类生存环境的关注。宋昆与胡子楠的作品《霾-天空》紧密围绕雾霾这一社会热点问题进行作品创作，以视频影像循环播放的方式展示了8年来作者自宅窗前同一位置同一时间段的图景，它们客观真实地记录了近年来我们身处城市的空气变化，上万只口罩围合出的空间界面，烘托出视频影像的氛围。该装置将空气、污染、雾霾这些抽象语汇进行符号化与图像化的处理，透过视觉震撼引发人们对未来城市生态环境的深刻反思。任军的作品《零亭》通过植物碳汇、再生能源以及循环材料的使用实现了一个水面上的零碳花园，在极度物质化的世界中塑造了一处聆听非物质社会中自然之音的场所。此外，该构筑物的所有材料全部来自"淘宝"，在某种意义上也探讨了建造之物在中国制造和"互联网+"背景下的新可能性。

本次展览，基于视点、视域的空间存在问题成为思考当代与非当代的另一出发点。赵劲松的作品《视点与观点》通过对命题进行提炼，洞悉其中隐含的共在关系，以空间为媒介，由人在装置中不同视点所获取的视觉映像体现多重空间关系的共在，进而引向对于事物以及人与事物关系的哲学思考。王振飞的作品《存在》运用镜像原理呈角度布置镜面，如同空间信息的收集器，在现实空间的"人"与镜像空间的"环境"之间建立对话关系，透过真实与幻象呈现当代与非当代的在场互动。

面对当代与非当代的话题，对于时间本身的哲学思考也是诠释该命题的重要切入点。张大昕的作品《轮回之境》以中国传统太极图式为原型，透过不同的材料之"镜"反射古今幻影，设置通行之"径"寓意循环往复,并最终指向时间之"境"演绎万物轮回,在无限的迁化与禅替中思考永恒的时间命题。张曙辉的作品《玛尼堆中》以藏传佛教文化中的玛尼堆为设计原型,百余块"漂浮"的玛尼石经四周镜面的多重反射在空间中不断延展,使观者犹如置身浩瀚的圣石之间,而玛尼堆在藏文化中恰好寓意着永恒,以永恒来回应每段时间的切片正是该作品诠释设计命题的独特视角。

对于"当代中的非当代"的解读甚至可以跨越时间与空间,在此基础上有关边界的思考成为该命题的衍生纬度。那日斯的作品《边界》利用不锈钢板片的折射将边界打破,使所处现实与板片中的幻象重叠,变化的时间与各异的空间被不断拼合重组,今昔、彼此之间被重塑为一种无界的状态,以此作为对设计命题的辩证解读。陈天泽与甄明扬的作品《造界》创建了一种全新的阅读方式,将其他参展建筑师的作品运用虚拟现实技术进行展示,将现实的"界"在虚拟现实中进行重塑,在当代语境下超越时间、空间,甚至是当代背景下的物质实体,创造了气象万千的无形之"界"。

2017天津国际设计周建筑展是近年来天津首次举办的以建筑师为创作群体的主题展览。"当代中的非当代"不仅是对当代的批判性思考,更体现了一种多元价值观的交融与共生。展览中建筑师们所呈现的观点和视角并非局限于建筑领域,而是通过建筑学语汇传递出对环境、时间和空间、真实和虚拟等内在问题的辩证思考。总体看来,本届展览将视野聚焦于当代社会景观,以多维度的视角审视持续变化且具有多面性的问题,就这点而言,本届展览无疑是一个良好的开端,也为今后的探索与实践奠定了重要的基础。

Beining Park in Tianjin witnessed the opening of the architecture exhibition of Tianjin International Design Week in May, 2017. This is the first independent theme exhibition with the architects as the creative subject since Tianjin International Design Week was held successively for four times. The curatorial team are from the School of Architecture, Tianjin University. According to the development situation of contemporary China, the team try to explore the challenges as well as opportunities for the discipline of architecture brought by issues such as tradition and future, regionalism and globalization, ecological environment and the "Internet plus" in the context of rapid urbanization based on the proposition — "Non-contemporaneity in Contemporaneity". With the participation of the young and middle-aged architects from Beijing and Tianjin, the exhibition showcased their works in forms of structures, installations and virtual reality, which revealed their individual standpoints as well as long-time reflections.

Ⅰ. The Origin of the Exhibition and the Process of the Curation

As a large-scale creative design exhibition in northern China, Tianjin International Design Week brings together high-quality design resources from China, Italy, Japan, Germany and the Netherlands, and makes outstanding contributions to promote the creative design industry in local region and improve the aesthetic awareness of the public. Since the year of 2014, the event has been successfully held for nine times, aiming to discuss the connection between design activities and hot topics such as daily behaviors as well as change of the times. The themes includes "Life with Art"," Memory and Dream"," Walking, Standing, Sitting and Lying Down", "The Future Is Now" and so on. Targeting decision makers, designers at home and abroad, students and common people, Tianjin International Design Week gradually forms a diversified paradigm with exhibitions, competitions, forums, workshops and seminars as the carrier, and builds an essential platform for the cultural exchange in Tianjin region.

Architects are an important group with keen insight into the era, the society, the city and the culture, and the independent exhibition with architects as the creative subject joined the Tianjin International Design Week for the first time in 2017. The group intended to respond to the theme of the design week by means of design strategies from the interpretative perspective of architecture.

The Chief Curator for the architecture exhibition of 2017 Tianjin International Design Week 2017 was Professor Song Kun from the School of Architecture, Tianjin University. And the team members were Zhao Jinsong, Zhang Xinnan, Yuan Sinan, Hu Zinan and Zhao Wei. The curatorial team began their preparations in October 2016. The initial work involved the formulation of exhibition theme and form as well as the list of architects to be invited. In November 2016, the theme of the Architects Design Exhibition—"Non-contemporaneity in Contemporaneity" was determined in correspondence to the theme of 2017 Tianjin International Design Week — "The Future Is Now". And 12 architects were invited to participate in the exhibition. Born in the late 1960s and the early 1980s, these architects from diversified fields covering state-owned architectural design institutes, private enterprises, independent firms and universities, could demonstrate the sensitivity and insight of the new-generation architects both in theory and practice. In December 2016, the participant architects visited the Beining Park in Tianjin for site survey and booth selection. In February 2017, the first plan seminar was held for the participant architcts in the School of Architecture, Tianjin University. And in the following month Mr. Masayuki Kurokawa, the general consultant for Tianjin International Design Week and a Japanese master in industrial and architecture design, met and communicated with the participant architects in reference to the content of the exhibition. In early April 2017, the second plan seminar was organized in Tianjin Xiangsi

Creative Industrial Park so as to confirm the plan of the exhibited works and construction scheme;. in late April 2017, the construction of the works began and the publicity activities started alongside. The exhibition was officially held from May 12th to 17th, 2017.

Ⅱ. From "The Future Is Now" to "Non-contemporaneity in Contemporaneity"

Themed on "The Future Is Now", 2017 Tianjin International Design Week intended to draw new design power in the connection and transformation process between the present and the future as well as the dialectical reflection. The "now" is a road that you stroll past, a glimpse of an image that passes in front of your eyes, a touch of coldness by the palm, and a beam of light converged on the body; being practical, objective and abundant around us, it proves its own existence in the form of "object". On the other hand, the "future" is an attempt of search in confusion, a second of reflection in chaos, a breath of silence in the noise, a touch of a dream in the dark; being vivid, subjective and embodied in our mind, it demonstrates a unique power via "ideas".

In view of this theme, the curators believed that "design" was the integration of "objects" and "ideas". On the one hand, it was a physical grasp of reality and substantial objects; while on the other hand, it was a metaphysical perception of the world and the ego. The "now" is an affirmation of the known, the awareness of the body present and a here-and-now existence; it is overwhelming as an existence of the "ego". The "future" is a sense of curiosity towards the unknown, an expectation of the image presentation and another realm from here to there; it is lingering on as an experience of the "superego". And "design" is thus a combination of the "ego" and the "superego", embracing and evoking human perception and imagination. Both the "now" and the "future" are the measurement of time since each present moment once exists in the future and each future moment is the next present moment. The so-called design always exists both in the state of mind at the moment and in the indication of the next moment. It's the reflection on "the time for life",and a kind of "continuity". People face the present with the future in their mind; and they prepare for the future with the present in their mind.

Based on the above discourse, the curatorial team designated the architecture exhibition as "Non-contemporaneity in Contemporaneity". In the architectural context, contemporaneity has always remained a topic worth deliberation and discussion. "Contemporaneity" may be interpreted as a reflection and criticism concerning the "now"; while "non-contemporaneity" embodies in a sense the traditional memories that become the "now" and the vision for tomorrow alluding to the "future". It is the sincere wish of the curatorial team to meld the slightly philosophical analysis with the current situation of urban development in China and to seek different solutions through the works of participatant architects.

Ⅲ. An Overview of the Exhibition

The architecture exhibition consisted of 10 works by 12 architects, and they were sited in two two-story traditional Chinese buildings and their in-between area in Beining Park, Tianjin. The site types included indoor rooms and outdoor courtyards of the traditional buildings as well as the water space between the two buildings. The architects started from various perspectives such as tradition, future, space, time and boundary in their creation and presented their multidimensional reflections and interpretations on the theme.

In response to the site, "Tradition and Contemporaneity" naturally becomes one of the essential interpretations concerning the design theme. *The Red Pavilion* by Gu Zhihong bases its design on a traditional pavilion in the courtyard. The gradual fragmentation of the pavilion shape renders a sharp contrast between the top and the ground and thus conveys various perplexities and reflections on human existence between the pure tranquility of tradition and capricious disturbance of the contemporary world. Xu Qiang's work *Time of Colonnade* introduces western classical spatial elements into traditional Chinese courtyards by means of reverse mapping of traditional materials with the unique attributes of contemporary materials so as to deliberately create a conflict and juxtaposition of the abstract and the concrete and initiate the dialogue between tradition and contemporaneity as well.

The theme of "Non-contemporaneity in Contemporaneity" is not only concerned about the tradition, but also oriented towards the future. In this architecture exhibition, the pondering over "the contemporary and the future" come more from the focus on the human living environment. *Haze-Sky*, created by Song Kun and Hu Zinan, centers on the hot issue of haze and displays the views by the window of the architect's own house at the same time and spot for eight years by the continuous playback of videos and photos. These recordings document in an objective and faithful manner the air changes in our city in recent years. The space interface formed with tens of thousands of masks is suited to the atmosphere of video images. The installation makes symbolic and graphical processing for air, pollution, haze and other abstract languages, resulting in a visual impact that evokes people's profound introspection on the future urban ecological environment. The work *Zero Pavilion* by Ren Jun is a zero carbon garden on the water made of plant carbon sink, renewable energy and recycled materials, a place to listen to the voice of nature in the immaterial society against the extremely materialized world. In addition, all the materials of the composition came from Taobao, and the new possibility in the context of "made in China" and "Internet +" is explored in a sense.

Another starting point for the theme of the exhibition is the spatial existence based on point of view and field of vision. In the work named *Viewpoint and Point of View*, the exhibition theme was distilled by the architect Zhao Jinsong so as to reveal the implied coexistence. Taking space as a medium, he demonstrates the visual images obtained by people from different viewpoints in the installation, which reflects the coexistence of multiple spatial relations, and thus leads to the philosophical contemplation of things and the relationship between humans and things. The work *Existence* is presented by Wang Zhenfei, in which mirrors are arranged with a certain angle in accordance with the image theory. The work stands as a collector of spatial information to establish a dialogue relationship between "people" in the real space and the "environment" of the mirrored space and to display the present interaction between the contemporary and non-contemporary via the reality and the illusion.

In face of the topic of contemporaneity and non-contemporaneity, the philosophical thinking about time is an important point of penetration to interpret the theme of the architecture exhibition—"Non-contemporaneity in Contemporaneity". The work *Reincarnation* by Zhang Daxin is an artistic creation based on the archetype of the traditional Chinese Tai Chi pattern, mirroring both ancient and contemporary illusions by means of different materials. A path is paved implying the repeated cycle ad infinitum and pointing to the final realm of time embodying reincarnation. In the process of infinite transformations, the eternal proposition of time is pondered. The design prototype for the work *In the Marnyi Heap* by Zhang Shuhui originates from the Marnyi stone pile in the Tibetan Buddhist culture. More than a hundred "floating" Marnyi stones constitute a continuous extension via the multiple reflections of the surrounding mirrors, putting the viewers in a sea of holy stones. And as the Marnyi stones symbolize eternity in the Tibetan culture, the unique perspective of the work to interpret the exhibition theme is to respond to each individual slice of time with eternity.

The decipherment of "Non-contemporaneity in Contemporaneity" even transcends time and space. Based on this, thinking about the boundary becomes the derivative dimension of the proposition. In the work "*Bounday*", the designer Na Risi uses the reflection of stainless steel plates to break the boundary so as to achieve the overlap between the present reality and the illusion mirrored on the steel plate. The changing time and different spaces are constantly pieced together and recombined. The past and the present as well as here and there are reshaped in a boundless state and presented as a dialectical interpretation of the exhibition theme. In the work *Creating Zone*, Chen Tianze and Zhen Mingyang create a brand-new way of reading by displaying the works of other participatant architects via virtual reality technology, reshaping the "zone" in real life with VR. By doing so, they go beyond time, space, and even physical entities in the contemporary context and render a vivid kaleidoscope of invisible "zone".

The architecture exhibition of 2017 Tianjin International Design Week is the first theme exhibition held in Tianjin with architects as the creative group in recent years. The theme— "Non-contemporaneity in Contemporaneity" is not only critical contemplation about the contemporary era, but also the integration and symbiosis of multiple values. The viewpoints and perspectives presented by the architects in the exhibition are not limited to the field of architecture, but to convey a dialectical thinking on internal issues involving environment, time and space, reality and virtuality. On the whole, this exhibition focuses on the contemporary social landscape, and examines the continuously changing and multifaceted problems from a multi-dimensional perspective. In this regard, this exhibition has undoubtedly marked a blazing start, and has also laid a significant foundation for the future exploration and practice.

对天津国际设计周执行主席李云飞的访谈
An Interview with Li Yunfei, the Executive Chairman of Tianjin International Design Week

受访者:李云飞
采访者:胡子楠
采访时间:2017年6月23日
采访地点:天津巷肆文化产业园

胡子楠(以下简称"胡"): 迄今,天津国际设计周已成功举办了四届,活动内容涵盖了主题展览、竞赛、实验班、论坛等众多方面。很高兴能邀请到您,请您谈谈作为总策展人,创办天津国际设计周的缘起及其四年来的发展历程。

李云飞(以下简称"李"): 天津是一个传统又开放的城市,特别具有地域特色,其国际化程度也越来越高。有一句话说,天津是"万国建筑博览会",而且天津的市民很淳朴,所以我觉得天津这座城市特别有魅力。我当时以市人大代表的身份连续5年向市政府提交天津成为联合国教科文组织的设计之都的申请,但是前提是天津必须有一个属于自己的活动,所以2013年我通过市人民代表大会向天津市政府申请举办天津国际设计周,将文化和创意相结合,打造天津文化品牌。

天津国际设计周从2014年开始举办,到现在已经四年了,第一年主题叫"随艺生活",然后是"记忆与梦想""行住坐卧",今年是"未来就是现在"。这四年来各个地方的设计师、策展人越来越多地认可天津国际设计周,我们希望,所展出的设计作品越来越国际化,越来越有国际的语言,也希望每个民族、每个国家所设计的东西都有本民族的特色和精神。

胡: 近年来,中国设计在世界备受关注,中国设计在威尼斯双年展、米兰三年展、巴黎蓬皮杜国家艺术和文化中心、荷兰建筑协会等重要的国际展览和学术机构不断亮相。与此同时,在国内,诸如北京国际设计周、上海设计之都活动周、深港城市/建筑双城双年展等也逐渐形成自己的影响力,不断将中国设计展示给世界。那么,与国内外其他设计展相比,天津国际设计周自己的特点和优势是什么?在设计周策划中,如何突出天津的地域性与天津设计的特质?

李: 我们也在考虑这个问题。我去过米兰国际设计周,感觉特别成功,那是全世界设计师的盛会。但是意大利的朋友跟我说,你知道它有多少年的历史吗?50年了。对于天津国际设计周,我们希望一步一个脚印向前走,形成自己的特色、特点,慢慢地让该活动越来越被关注。现在有一个校际联盟,意大利教育部和中国教育部特别支持这个项目。目前意大利有几所大学和天津大学、南开大学及天津美术学院组成了一个校际联盟,学生之间可以进行交流,学校之间也会进行合作,而且这个联盟会扩展到越来越多的学校,全世界的学校都有可能加入。另外,教育也是我们特别重视的,我们做了一个培训中心,这个培训中心目前在天津有意大利语等级考试中心。我们也创办了一个艺术类的高中,这是在设计周的概念下做的,这个高中类似于一个预科班,学生们在这里学习英语、意大利语、法语,以便于出国学习。同时,政府希望了解这些学生在国外的学习状态,跟踪他们在国外的学习进度,因此学生们需要每年做一次汇报。政府希望他们特别努力地学习,能代表中国年轻人的形象,将来能回国创业。

另外,我们的竞赛越来越注重实用性,希望能将他们的设计变成产品,这也将成为天津国际设计周的一个亮点。同时,我们做了一个设计博物馆,让更多的年轻人了解国外产品的品质,希望学生能够近距离接触这些作品,了解其在设计与制作工艺上的特质。这个设计博物馆未来也会成为天津市的特色。

胡: 过往四届设计周的主题分别为"随艺生活""记忆与梦想""行住坐卧"以及"未来就是现在",涉及人们生活模式的变化、技术的发展以及人类思想的革新。从类型上来说,前3年是在关注人的日常生活和行为,今年发生了一个转变,开始关注时代发展的问题,特别是工匠、环境、科技、经济等主题的引入,使得本次设计周更具社会性的一面。这样的主题发展线索是基于什么观察与思考的?是否意味着天津国际设计周的举办思路与当下社会发展的联系更加紧密了,向着开放性与多元化发展了呢?

李: 天津国际设计周今年的主题是"未来就是现在",是由意大利团队提出的,它秉持了"设计无处不在"的理念,旨在对社会、经济等问题进行全面思考。从展览上看,今年设计周更多地融入了科技、工匠、现代化的元素,尤其是德国的展览。手工艺在德国具有特殊的地位,汽车、机器、啤酒等德国制造已被我们熟识。在天津,工匠同样扮演着十分重要的角色,设计周所展示的工艺品均来自大师级的艺术家、设计师和手工匠人,在他们的细心雕琢和对完美细节的追求下,这些作品被赋予了更独特的价值。天津国际设计周未来的发展方向必然是顺应当下社会的发展规律,更加开放和多元,进一步推进"天津制造"向"天津创造"的转化。

胡: 在现代建筑百余年的历史中,各类展览是新锐建筑师活动的重要舞台,展览起到了引领先锋的作用,甚至促进了新建筑流派的产生。以建筑本体为中心的展览是20世纪一种独特的现象,如1925年巴黎国际现代化工业装饰艺术展览会、1927年斯图加特魏森霍夫建筑展、1969年在纽约现代艺术博物馆举办"纽约五"建筑展、1988年纽约现代艺术博物馆的"解构主义建筑"7人作品展。很多展览与现代建筑史密不可分，因此成为其重要的组成部分。今年天津国际设计周首次增设了建筑展的环节，以天津地区为主的青年建筑师的作品亮相设计周，请您评价一下这次青年建筑师的设计展。作为一个全新的环节，它将对天津地区的建筑设计活动带来哪些方面的推动力量？

李： 在本届天津国际设计周中，建筑师们也加入对"当代"的探讨中，形式感极强的结构线条和材质实验，俨然谱就了一首建筑师们的狂想曲，这12位建筑师的10件作品体现了天津建筑设计的水平。我们跟很多设计周的区别是有一个相对固定的展馆，天津大学、南开大学、天津美术学院都有分会场，还有一些其他展览馆，虽然数量不是特别多，但是我们希望品质有保证，展览每年都会有一些增长，未来我们也希望展览像米兰设计周一样，全城都能参与这个活动。但是我们也要谨慎，不要使其变成一个既缺少学术性又缺少内涵的旅游项目。

胡： 建筑展览作为当代建筑文化的巨大推进器，应当承担更重要的使命。从理论角度而言，建筑展应作出学术贡献，并带来社会效应。一方面建筑展可以梳理或挖掘一些建筑理论线索，主动收集相关建筑师的作品，并按主题顺序展出，以展览的形式剖析当代建筑设计思潮的发展动向，以引领建筑设计实践界内的发展趋势；另一方面，具有较高学术价值的建筑展通常关注技术发展或者文化变迁在建筑设计实践中所体现的变化。展览将某个学术的命题展现在人们面前，引起对该学术命题的讨论和争辩，从而推动建筑学术研究的发展。从您的角度看，您对于天津国际设计周建筑展的定位是什么？它未来的发展模式和方向是什么？

李： 首先，天津国际设计周可以成为一个中国设计师交流的平台，也可以成为一个中国和国外设计师交流的平台，甚至可以成为一个大学之间、建筑师之间、设计师之间、厂商和设计师之间交流的平台，为大家带来一场文化的盛宴，让大家了解设计与生活之间的关系。其次，我们的市民有一些对美和艺术的看法，这或许是无意识的，在买家具或者选择某种生活方式时，他们的选择不是出于生活本身和艺术的角度，而是出于其他原因，我们希望他们对美的概念有清晰的认识之后再选择自己的生活方式，我觉得这也是第二个非常重要的方向。

今年我们增加了建筑展，为天津国际设计周增添了新意。现在有一个概念叫"永不落幕的设计周"，我希望在未来，这个建筑展可以代表天津成为自己的一个品牌，通过在全国做巡展，让更多的人了解天津，知道天津国际设计周。

胡： 今年的建筑展是在两个"四合院"中完成的，有趣之处在于，这是在一个"传统"的场所语境中，通过一些建筑装置展开对一些"当代"问题的思考，对于明年的建筑展，在策展方式、场地选择、展览模式上您有什么展望？

李： 明年设计周的时间定在2018年5月11—16日，主题是"东方×西方"。我们每年的主题都不同。我们关注设计周的团队并希望这些团队继续努力，一直有一个清晰的想法往前走。如果每年不停地换人，你会发现每个想法都很精彩，但是却失去了特色，我们也一直在调整，比如将今年的主题、明年的主题、未来的主题串起来就变成了天津国际设计周的特色，我们还是希望能打造出自己的特色。建筑展也是一样，明年会继续在北宁公园举行，在有一百多年历史的中国古典园林里，做特别现代的活动，我们觉得这特别有文化价值。明年的建筑展我打算在一个废弃的仓库里面举行，可以让设计师尽情地发挥想象力。我相信明年的建筑展上建筑师们会有更加精彩的表现。

Interviewee: Li Yunfei
Interviewer: Hu Zinan
Time of Interview: 23rd June, 2017
Place of Interview: Tianjin Xiangsi Creative Industrial Park

Hu Zinan (hereinafter referred to as "Hu"): So far, Tianjin International Design Week has been successfully held for four sessions. The event involves various activities such as theme exhibitions, competitions, experimental classes, forums and so on. It is a great pleasure to invite you. As the chief curator, please talk about the origin and development of Tianjin International Design Week in the past four years.

Li Yunfei (hereinafter referred to as "Li"): Tianjin is a traditional and open city with special regional characteristics and ever-increasing internationalization. There is a saying that Tianjin is known as the "International Architectural Exposition", and people here are very simple and honest. So, in my eyes, the city is particularly charming. As a member of the Municipal People's Congress, I submitted for five consecutive years to the Municipal Government of Tianjin the application that Tianjin should be elected the UNESCO City of Design. But the premise is that Tianjin must host an event of its own. As a result, I applied in 2013 to the Municipal Government of Tianjin via the Municipal People's Congress to host the Tianjin International Design Week in an attempt to combine culture and creativity and build a Tianjin cultural brand.

It has been four years since the first Tianjin International Design Week was held in 2014. The theme for the first year is "Life with Art", the second year is "Memory and Dream", and the third year is "Walking, Standing, Sitting and Lying Down", and this year is "The Future Is Now". The past four years has witnessed a growing recognition of our exhibition by designers and curators from all parts of the world. And it is our sincere wish that the design works on display are more and more internationalized using international languages, and that every ethnic group and every country maintain their own characteristics and spirit in design.

Hu: In recent years, Chinese design has attracted much attention in the world. Chinese design has been showing up in important international exhibitions and academic institutions such as the Venice Biennale, the Milan Triennial, the Pompidou Centre in Paris, and Netherlands Architecture Institute. At the same time, domestic events such as Beijing International Design Week, Shanghai Design Week, and Shenzhen-Hong Kong Bi-City Biennale of Urbanis\Architecture (UABB) have also gradually formed their own influence, and constantly display Chinese design to the world. So, compared with other design exhibitions at home and abroad, what are the characteristics and advantages of Tianjin International Design Week? In the planning for the design week, how could you highlight the regional features and the design characteristics of the city?

Li: What you asked just now is what we are thinking about. I went to Milan International Design Week, which struck me as a very successful exhibition and a grand event for designers all over the world. But one Italian friend told me that the exhibition has a history of 50 years. As for the Tianjin International Design Week we intend to make progress one step at a time, endowing it with its identity and characteristics. By so doing, the event will have a better chance of getting noticed. We are now hosting an Intercollegiate League that has won special support from both Italian Ministry of Education and the Chinese Ministry of Education. This league is composed of several Italian universities and some universities in Tianjin such as Tianjin University, Nankai University and Tianjin Academy of Fine Arts. This inter-school league is committed to the communication between students and cooperation between universities. And in the future the league will expand to include more and more schools around the globe. Furthermore, we attach special importance to education. We built a training center, which currently has an Italian language level examination center in Tianjin. In addition, we run an arts-based high school, which originates from the concept of the Design Week. This high school is similar to a preparatory class, through which high school students study foreign languages such as English, Italian and French in order to study abroad. At the same time, the government is concerned about the study of these students abroad, so an annual report is needed to submit to keep track of their learning abroad. And these young people are expected to represent Chinese young people, study hard, and return to their homeland to start their own businesses in the future.

In addition, our competition is increasingly laying emphasis on practical purposes, with the hope of converting the designs into actual products, which will also become one highlight of Tianjin International Design Week. Meanwhile, a design museum was established to introduce foreign products to young students, offering them a close access to the design and production technology of these works. This design museum will also become a specialty of Tianjin in the future.

Hu: The themes for the previous four design weeks are "Life with Art", "Memory and Dream", "Walking, Standing, Sitting and Lying Down", and "The Future Is Now" respectively, involving the change of people's life pattern, the development of technology and the innovation of human thoughts. As far as the type is concerned, in the first three years, the themes of the exhibition deal with people's daily life and behavior while this year there is a shift towards the development of the times, especially the introduction of themes such as craftsmanship, the environment, science and technology, economy and etc, which provides this design week with more social traits. What observations and contemplations are such theme-developing clues based on? Does it mean that the notion of Tianjin International Design Week is more closely related to the current social development and moves towards openness and diversification?

Li: The theme of Tianjin International Design Week this year is "The Future Is Now" proposed by Italian team, which upholds the concept of "Design is omnipresent" and aims to consider social and economic problems in an all-round way. As for the exhibition, more elements such as technology, craftsman and modernization are integrated into the design week this year, the German exhibition being a case in point. Handicraft art occupies a unique position in Germany, as we are familiar with cars, machines, beer and other goods manufactured in German. Here in Tianjin, craftsmen also play a very important role, and the handicrafts on display are all from master artists, designers and craftsmen. These works are endowed with peculiar values owe to their refinement and the pursuit of perfect details. The future development of Tianjin International Design Week is bound to respect the law of development of the current society, more open and diversified, and further propel the transformation of "Made in Tianjin" towards "Created in Tianjin".

Hu: In the history of modern architecture with more than one hundred years, various kinds of exhibitions are main stages for the avant-garde architects. These events have played a leading role and even promoted the emergence of new architectural schools. Exhibitions centering on architecture itself are a unique phenomenon in the 20th century, such as the International Exposition of Modern, Industrial and Decorative Art in Paris in 1925, the Stuttgart Weissenhof Estate in 1927, the New York Five at the Museum of Modern Art (MOMA) in 1969, and the 7-architect work exhibition entitled "Deconstructivist Exhibition" at MOMA in 1988. Many exhibitions are inseparable from the history of modern architecture and thus become an essential part of it. This year, a new component of Architecture Exhibition is added for the first time to Tianjin International Design Week. The works of young architects primarily from Tianjin make their debut in the design week. Please make some comments on this design exhibition of young architects. As a brand-new component, what advancement will it bring to the architectural design activities in Tianjin?

Li: During this Tianjin International Design Week, architects also joined in the discussion of "contemporary". The structural lines and material experiments with an intense sense of form are like a piece of rhapsody composed by these architects. Ten works by twelve architects embody the level of architectural design in Tianjin. We differ from many other design week exhibitions in that we possess a relatively fixed exhibition hall with branches in Tianjin University, Nankai University and Tianjin Academy of Fine Arts as well as exhibition halls in other places. Though this is not a large number, we still strive for high quality and annual increase in the number of exhibitions. In the future we also envision the participation of the event at the city-wide level like Milan Design Week. But we should also take caution and avoid the tendency of rendering a tourism project which lacks both academic traits and cultural content.

Hu: As an immense propeller of contemporary architectural culture, the Architecture Exhibition is obliged to assume a more important mission, making academic contribution as well as bringing social effect from a theoretical perspective. On the one hand, some theoretical clues may be sorted out or revealed in these exhibitions by actively collecting the works of relevant architects and putting them on display according to their themes, and we are capable of exploring the development route of contemporary architectural design concepts in the form of exhibitions so as to lead the development trend in the field of architectural design practice. On the other hand, Architecture Exhibitions with relatively high academic values usually focus on the changes in architectural design activities reflected by technological advancement or cultural changes. The exhibition presents an academic proposition to the public, causing a discussion and debate and thus promoting the progress of academic research in architecture. From your point of view, what is your positioning for the Architecture Exhibition of Tianjin International Design Week? And what is its future development pattern and direction?

Li: In the first place, Tianjin International

Design Week can be a platform of communication not only among Chinese designers but also between Chinese and foreign designers, and even among universities, architects, designers, manufacturers and designers, which can bring a cultural feast to the public and make people understand the relationship between design and life. Secondly, our citizens, out of their unconsciousness, may have some viewpoints concerning beauty and art. When buying furniture or choosing a certain way of life, they tend to make the choices not based on the perspective of life itself or art, but for other reasons. We hope that they won't make the choice until they are clearly aware of the concept of beauty. I believe this is another important direction.

We added Architecture Exhibition this year, injecting some originality into Tianjin International Design Week. Now there is a concept "Never-ending Design Week". I envision that in the future, this Architecture Exhibition can become a brand of its own on behalf of Tianjin, by means of tour exhibitions across the country, make more people know about Tianjin, and about Tianjin International Design Week.

Hu: The Architecture Exhibition was completed in two "quadrangle courtyards" this year. What is interesting is that some studies on "contemporary" problems are carried out via some architectural structures in the context of a "traditional" site. As for next year's Architecture Exhibition, what expectations do you have in relation to the curatorial pattern, site selection and exhibition mode?

Li: The Design Week for next year is set on May 11th to 16th, 2018, with the theme— "East×West". The theme for the exhibition every year tends to vary. We focus on the teams in the Design Week and hope these teams will keep on with their efforts and forge ahead with an explicit idea. If you keep changing people every year, you'll find that every idea is great, but it loses its character. Adjustments have been made accordingly and the series of themes for this year and the following years are linked to be a feature of Tianjin International Design Week. We still strive for distinctiveness in our exhibition. The same is true for the Architecture Exhibition as well. Next year, the exhibition, as a remarkably modern event, will still take place in the Beining Park, a Chinese classical-style garden with a history of over 100 years. And we consider it an event of high cultural value. I intend to hold the Architecture Exhibition for the coming year in a deserted warehouse, which allows designers to give full play to their imagination. And I believe the architects will have more brilliant performance in the Architecture Exhibition next year.

2017 参展建筑师及作品
Participant Architects & Their Works

陈天泽
甄明扬
Chen Tianze
Zhen Mingyang

天津市建筑设计研究院有限公司
Tianjin Architecture Design Institute Co., Ltd.

造界

"须弥藏芥子,芥子纳须弥"是中国古代的一句禅语,说明即使是极小的物体也可以容纳万物。这并非物体本身所具有的特征,之所以这么讲,是因为当物体拥有意识时,意识的空间可以是无限的。当意识脱离现实的秩序,即使是最小的空间也可以容纳最大的内容。这个最小的空间,就是创建的"界"。

该装置借助当代VR技术,结合古代禅学思想及对未来意识空间的畅想,在最有限的空间中造一个"界",试图让观众从中体验一个虚拟的设计周,脱离现实的束缚,让设计师最原初的思想与观众直接碰撞。同时,通过虚拟世界中原始设计与实际建成效果的对比,观众可以体验到建筑创作中的时间属性。此装置不仅给予观众现实展览之外新的思考,也为设计周主题"当代中的非当代"提供另一种解读视角。

Creating Zone

There is a Zen saying that the world contains every smallest object, but every smallest objects can keep a countless world inside. It is not about the physical feature of the object, but for the reason that the consciousness space will be infinite when the object can have its own consciousness. When consciousness can be detached from the physical order, even the smallest space can accommodate the largest content. And this smallest space is the "zone" wichi is created.

With the help of VR technology, according to ancient Zen thought and the imagination of the consciousness space in the future, the installation creates a "zone" in the most limited space. We are trying to let the audience experience the design week in this "zone", avoiding the shackles of reality. In this way, the designer's thoughts could be directly showed to the audience. At the same time, through the comparison of the original design and the actual completed effect in the virtual world, the audience can experience the time attribute of architectural design which can help the audience to think beyond the exhibition. And the "zone" can also provide an explanation perspective for the theme of the design week—"Non-contemporaneity in Contemporaneity".

顾志宏
Gu Zhihong

天津大学建筑设计规划研究总院有限公司/顾志宏工作室
Tianjin University Research Institute of Architectural Design and Urban Planning Co., Ltd./Gu Zhihong Studio

红亭子

"红亭子"是4万个8cm×8cm闪光的红色方块与北宁公园既有的传统风格的红亭子的整合。该建筑上半部分仍保持传统风格,下半部分则用红色方块缠绕四根亭柱并将其延伸至地面。伴随着亭柱上红色方块的抽象演化,地面上一堆闪光的碎片就像是历史的倒影。不管是传统还是未来,都与当代产生了有趣的矛盾,这也是当代中国建筑师的困惑、挣扎与思考的直接反映。继承和发扬中国优秀的传统文化,是中国建筑师的职责和担当。和这个亭子一样,现实、历史与未来都存在着延续、发展和演化的自然过程,同时又常存在着矛盾和剧变。这些闪光的碎片和代表中国传统文化的红亭子紧密地融合成一个整体,就像中国文化深深地植根于我们的血液中一样。

通过审视这个红亭子,不难想象和思考这个问题:这些闪光的碎片折射出当代快节奏的、多变的、闪光的生活,在这个快节奏的时代,该如何传承和发展历史传统文化?这是一个重要的课题。"红亭子"揭示了一种困惑和思考。"红亭子"的整体性和矛盾性间接表达出对建筑师继承传统、面对未来的困惑和反思。

Red Pavilion

"Red Pavilion" is the integration of forty thousand 8cm×8cm shiny red squares with an existing traditional red pavilion in Beining Park. The top half of the pavilion still keeps the traditional style; the lower half uses red squares to twist around four columns and extends them to the ground. With the abstract evolution of red squares on the columns, a pile of shining fragments on the ground are just like the reflection of the history. No matter it is tradition or future, both of them have some interesting contradictions with the contemporary era, which is also a direct reflection of Chinese contemporary architects' confusion, struggle and thoughts. To inherit and promote Chinese traditional culture is the responsibility of Chinese architects. Like this pavilion, all of the present, the history and the future have a natural process of continuation, development and evolution, and there are always contradictions and great changes at the same time. These shining fragments and the red pavilion which represents Chinese traditional culture are closely integrated into a whole, just as Chinese culture is deeply rooted in our blood.

By inspecting this red pavilion, it is not hard for us to imagine and think about this question: these shining fragments reflect the contemporary fast-paced, variable and shining life, and then how to inherit and promote the historical and traditional culture in this fast-paced age? This is an important issue. And "Red Pavilion" reveals a confusion and consideration. The integrity and contradiction of "Red Pavilion" indirectly expresses the confusion and reflection on how to inherit the tradition and how to face the future.

那日斯
Na Risi

天津市博风建筑工程设计有限公司
Tianjin BF Architecture Design Co., Ltd.

边界

"当代中的非当代"应该是当代人思维下的一个跨越时空的持续性空间，而非短期流行性的产物。"边界"这件作品的意义就在于打破这一界限，通过材料的特性和空间变化的过程引发人们思考。

边界是一种状态，处于这边和那边的中间，是这边与那边彼此的渗透和印证。水平、垂直、悬浮、映射，表现的是一种边界和无边界的空间状态。在穿梭和驻足的过程中，人们寻求多重的空间可能性。真实与虚幻的重叠，好像世间万物不断变化的过去、现在与未来，而这一切实际上是无边界的。

设计主要由两部分组成，一部分是中间区域的上、下两个大镜片，另外一部分是围绕大镜片悬挂的条状镜片。条状镜片从展区入口处，通过由疏到密的排列，将人慢慢引入一个狭窄的空间，最后进入无边界的立体空间。在行走的过程中，人们每走一步，伴随着水滴声和光效，周围环境都会发生变化。形、声、闻、触四感为人们带来一种多维度的体验。

材料主要选用304双镜面不锈钢板。不锈钢板被切割成不同的尺寸，从屋顶垂吊而下。人们实际穿梭垂吊的镜面时产生的声音和我们原本营造的水滴声会产生两种不同的效果，一种如打雷般轰鸣，另一种静谧如水。垂吊着的镜面在灯光下犹如波光粼粼的水面，配合镜片倒映在地面上的斑斓光影，显现出别样的美。这些设计之外的惊喜，更体现出设计本身所要表达的多重可能性。

"边界"在"当代中的非当代"中所要呈现的意义更多的是空间延伸的表达，从二维、三维的世界发展到更多维度。从繁杂的现实社会中脱离一分钟，进行头脑风暴，在波光下寻找自己，相视而笑，然后消失在边界中，不失为一种特别的体验。

Boundary

"Non-contemporaneity in Contemporaneity" should be a continuous space spanning time and space under the thinking of contemporary people, rather than a product with short-term popularity. The significance of this piece of work "Boundary" is to break this boundary and to trigger people's thinking through the characteristics of materials and the process of spatial change.

Boundary is a state between this side and that side, which is the mutual infiltration and confirmation of this side and that side. Horizontal, vertical, suspension and mapping express a spatial state between boundary and no boundary. In the process of shuttle and pause, people seek multiple spatial possibilities. The overlap of truth and unreality is like the ever-changing past, present and future of all things in the world, and in fact all these are boundless.

The design is mainly divided into two parts, one part is made up of two large lenses in the middle area, and the other part consists of strip lenses hanging around the large lenses. From the entrance of the exhibition area, the strip lenses are arranged from sparse to dense, slowly introducing people into a narrow space, and finally into a borderless three-dimensional space. In the process of walking, the sound of water droplets and light effects make the surrounding environment change when people take every step. The four senses of form, sound, smell and touch bring people a multi-dimensional experience.

The main material is 304 double mirror stainless steel plate. The stainless steel plates are cut into different sizes and hung down from the roof. The sound produced when people actually shuttle the hanging lenses and the sound of water droplets created by ourselves, create two different feelings. One kind is thunder-like roar, and the other kind is quiet like water. The hanging lenses resembles a shimmering water surface under the light, together with the brilliance of the light and shadow reflected on the ground by the lenses, creating a different kind of beauty. These surprises also reflect the multiple possibilities of the design itself.

The meaning of "Boundary" in the "Non-contemporaneity in Contemporaneity" is mostly the expression of spatial extension, which develops from the two-dimensional and three-dimensional world to more dimensions. It is a special experience to break away from the complex real society for a minute to brainstorm, look for ourselves under the light, smile at each other, and then disappear into the boundary.

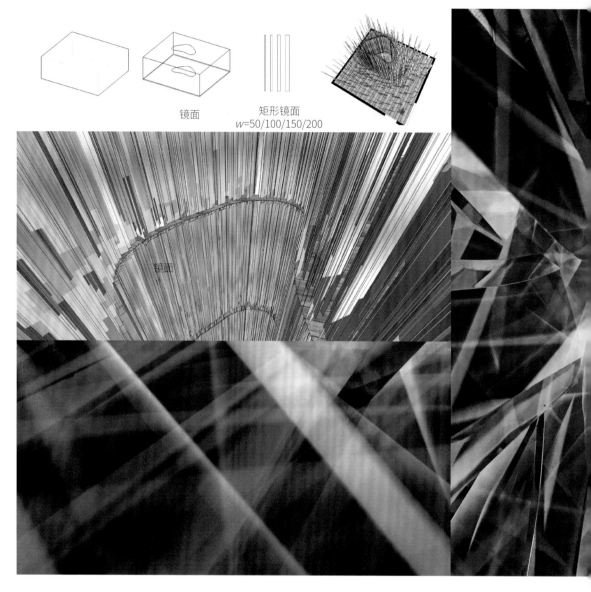

任军
Ren Jun

天津市天友建筑设计股份有限公司
Tianjin Tenio Architecture and Engineering Co., Ltd.

零亭(音同"聆听")—— 一个"淘宝制造"的"零碳花园"
针对建筑展的主题"当代中的非当代",该设计以一组绿色装置探讨时空线索中的当代生活。此装置是一个位于水面上的零碳花园——零亭——聆听非物质社会中自然的声音。

当代中的非当代——聆听时空
从减少碳排放的生态角度出发,作为临时性展览装置的零碳花园具有轻质(不做重型基础)、透明(仿佛不存在)、漂浮(轻巧地悬在水面)、循环(循环材料循环利用)、人本(人可以使用它)的特点,同时可以与植物和水一起融入公园中。

零碳花园
零碳花园的实现途径首先是利用植物碳汇固碳释氧,其次是布置可再生能源抵消建造过程中的碳排放,同时尽可能采用废弃或循环材料以减少材料的使用,最终实现一个碳中和的景观花园。

淘宝制造
零亭的所有材料,除了亭子顶上北宁公园的树叶和花瓣,从聚碳酸酯(PC)波形板到太阳能灯,从连接螺栓到空气凤梨全部来自淘宝。

也许可用一首诗歌来描述这个小小的临时装置。

零亭
一个亭子漂浮在水面上,轻盈剔透到仿佛不存在。
存在的是悬空的松萝,用绿色画出唐伯虎的水墨。

从"Made in China"到"Made in Taobao",
现实和虚拟,被链接在无所不包的网络帝国。
碳汇的植物长在亭中,长在水上,长在岸边。
柔软薄膜引太阳发电,借回收的旧物串起零碳花园。
历史中的前自然生态,我们寻找当代中的非当代。
未来中的后工业虚拟,时空中也许未来就是现在。

Zero Pavilion (Chinese Pronunciation Similar to "Listening")—A Zero Carbon Garden Made by Taobao
In view of the theme "Non-contemporaneity in Contemporaneity" of Architecture Exhibition, I hope to explore contemporary life in space-time clues with a group of green devices. Our installation is a zero carbon garden on the water — Zero Pavilion — listening to the voice of nature in the immaterial society.

Non-contemporaneity in Contemporaneity— Listening to Time and Space
From the ecological perspective of reducing carbon emission, we hope that, as a temporary exhibition device, this zero carbon garden has the characteristics of light weight (no heavy foundation), transparency (as if it didn't exist), floating (hanging on the water lightly), recycling (recycling the recycled materials), and people-oriented (people can use it). At the same time, it can be integrated into the park with plants and water.

Zero Carbon Garden
The first way to achieve zero carbon garden is to use plant carbon sink to fix carbon and release oxygen. The second way is to arrange renewable energy to offset the carbon emission in the construction process. At the same time, waste or recycled materials are used to reduce the use of materials as much as possible, so as to achieve a carbon neutral landscape garden.

Made in Taobao
All the materials of Zero Pavilion except the leaves and petals on the top of the pavilion that came from Beining park, from polycarbonate (PC) corrugated boards to solar lamps, from connecting bolts to air pineapples —came from Taobao.

Maybe a poem can be used to describe this little temporary installation.

ZERO PAVILION
A pavilion floating on the water, with light and transparent materials as if

nothing was there. Tillandsia levitate in the air, Draw a Tang Bohu's painting by green color.

From "Made in China" to "Made in Taobao", reality and virtuality are linked into all-embracing Internet empire.
Carbon sequestration plants grow in the pavilion, on the water, on the shore.
Soft films provide solar energy, and together with recycled old things form Zero Carbon Garden.
From the former natural ecology of the history, we are looking for the non-contemporaneity in contemporaneity.
According to the post-industry simulation in the future, maybe the future is now in space -time.

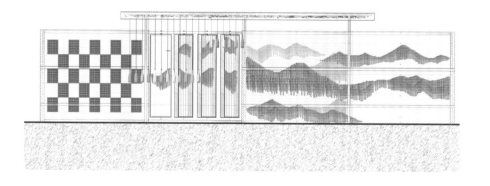

宋昆
胡子楠
Song Kun
Hu Zinan

天津大学建筑学院
School of Architecture, Tianjin University

霾——天空

在6200mm×6600mm×3000mm的空间中,四周墙壁与屋顶被白色口罩所覆盖,形成一个"界面",其中一面墙壁上悬挂屏幕,象征家宅窗户,并在其上播放"影像"。此设计通过"影像"和"界面"这两种元素来回应本次建筑展的主题——"当代中的非当代"。

影像——时间、空气与城市

对于时间的思考,最简单的方式便是将其记录下来。作品中播放的影像是由1644张照片连续播放形成的,这些照片记录了近79个月来每日中午12:00—14:00作者自宅窗前的景象。在这近8年的时间里,窗外的世界发生了很大变化,一方面是城市轮廓线的变化,高层建筑拔地而起,城市的天际线被不断修改,绿色的空间不断减少;另一方面与我国的生态环境相关,2017年以前每年雾霾天数越来越多,程度越来越重,城市被淹没于雾霾之下,这种视觉状态完全消隐了场所和人们的存在感,身处当代的我们必须警觉此现象。

界面——日常、手工与大众

口罩是将雾霾与人们的生活有效联系在一起的日常之物。在当代环境问题凸显的背景下,口罩作为一种简易的、随身携带的空气过滤器,基本成为空气污染、雾霾等一系列概念的实体符号。在这一作品中,口罩代替了常规的建筑材料,成为构筑建筑饰面的唯一元素,在产生一种意料之外的空间感受的同时,还期望以日常之物唤起人们对环境问题的迫切关注。

Smog—Sky

In the space of 6,200mm×6,600mm× 3,000mm, white masks spread over the surrounding walls and the roof to form an "interface". The window of the room is symbolized by a screen hanging on one of the walls, playing images. The two elements of "images" and "interface" are used to respond to the theme of this Architecture Exhibition— "Non-contemporaneity in Contemporaneity".

Image—Time, Air and the City

The easiest way to reflect on time is to record it. In this work, the images played on the screen are formed by the continuous playback of 1, 644 photos, which document the scenes outside the author's window from 12:00 to 14:00 every day for nearly 79 months. In the nearly eight years, the world outside the window has changed greatly. On the one hand, the city's outline has been changing, high-rise buildings have been appearing, the skyline of the city has been continuously modified, and green spaces have been shrinking. On the other hand, the environment is getting worse before 2017, the number of smog days increases, and the smog's degree is getting severe. Gradually, the city is being submerged under the smog. This visual state completely obscures the existence of places and people. This is a warning sign that we must be alert in contemporary times.

Interface— Daily Life, Handcraft and the Public

Mask is an object in daily life that effectively connects smog and people. As a simple and portable air filter, mask has become a physical symbol of a series of concepts such as air pollution, smog in the context of prominent environmental problems in the contemporary era. In this artwork, masks have replaced conventional building materials and become the only element for architectural finishing. While producing an unexpected spatial feeling, we still hope everyday objects can arouse people's urgent attention to environmental issues.

王振飞
Wang Zhenfei

HHDFUN事务所
HHDFUN Architects

存在

展览的场地位于中国传统花园中的一座建筑内,如何将室外的景观引入室内成为该作品的核心。

"存在"是为这次展览定制的作品,共三件,以不同的形态布置在三个房间中。作品本身由两面成一定角度的镜子构成,适当的折叠角度使得观者正对镜子的时候看不到自己在镜子中的虚像,不过由于镜面的不同角度以及在不同房间中的特定位置,室外的园林景观得以反射至室内,观者可以观察到不同景框中的借景。虽然观者看不到自己在镜中的虚像,但是站在不同角度的其他观者却可以看到前者的虚像融入借景当中。

Existence

The venue of the exhibition is located in a building in a Chinese traditional garden. How to introduce the outdoor landscape into the interior becomes the core of the work.

The work "Existence" is customized for this exhibition. There are three pieces in total, which are arranged in three rooms in different forms. The work itself is composed of two mirrors with a certain angle. The proper folding angle prevents the viewer from seeing his virtual image in the mirror when he is facing the mirror. At the same time, due to different angles of the mirrors and specific locations in different rooms, the outdoor garden landscape can be reflected indoors, and the viewer can observe the borrowed scenes in different frames. Although the viewer cannot see his virtual image in the mirror, other viewers from different angles can see the former's virtual image blended into the borrowed scene.

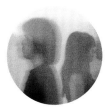

徐强
Xu Qiang

天津天华建筑设计有限公司
Tianjin Tianhua Architecture Planning & Engineering Ltd.

时间的柱廊

场地是一栋小别墅，本身就带有中式古典语言。别墅的院落从尺度上看是一个长的线性空间；从功能上看是个通过性空间；从感受上看是个围合性空间。四面为青砖围墙，墙上有砖雕和镂空的窗洞，空间本身就带有很强的非当代性，而且是东方的非当代性。在这个东方的非当代空间中，该装置营造了一种冲突感——把东方的、西方的，现在的、过去的，具象的、抽象的元素全部放置在一起，用这种矛盾冲突使人们体会到"当代中的非当代"。

东方的典型古典空间是院落，西方的典型古典空间是柱廊。首先，一个抽象的柱廊空间被植入场地，这就形成东方与西方的并置。其次，柔性的聚氯乙烯（PVC）膜被用来模拟柱廊、拱券，用柔软体现坚硬，用轻盈体现沉重，用色彩体现灰度，一个抽象的逐浪空间即被植入一个青砖围合的院子中，这就是具象与抽象的并置。院子代表过去，柱廊代表过去，装置材料又代表现在和未来，这即是现在与过去的并置。

Time of Colonnade

The site is a small villa with classical Chinese elements. The villa's courtyard is a long linear space from the perspective of scale; in terms of function, it is a transitive space; in terms of experience, it is an enclosed space, surrounded by blue brick walls with brick carvings and hollow windows on the walls. The space itself has a strong non-contemporary character, and what's more, it is the non-contemporary character of the East. In this eastern non-contemporary space, the installation creates a sense of conflict, all elements are put in the East and the West, in the present and the past, concrete and abstract, and use this contradiction and conflict to make people feel "Non-contemporaneity in Contemporaneity".

The typical classical space in the East is the courtyard, and that in the West is the colonnade. First of all, an abstract colonnade space is implanted into the selected place, which forms the juxtaposition of the East and the West. Secondly, the flexible polyvinyl chloride (PVC) film is used to simulate colonnades and arches. The softness reflects the hardness, the lightness reflects the heaviness, and the color reflects the grayscale. And an abstract wave-riding space is embedded in a blue brick enclosure, which is the juxtaposition of concreteness and abstraction. The courtyard represents the past, the colonnade represents the past, while the installation material represents the present and the future, which is the juxtaposition of the present and the past.

抽象 变异

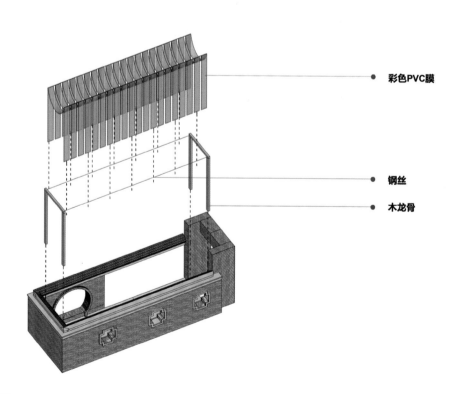

彩色PVC膜

钢丝

木龙骨

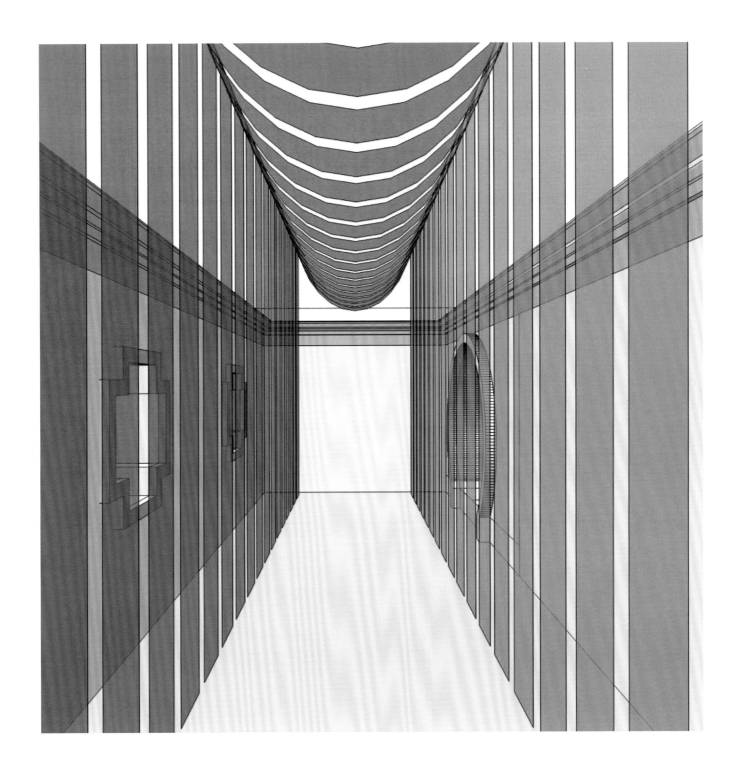

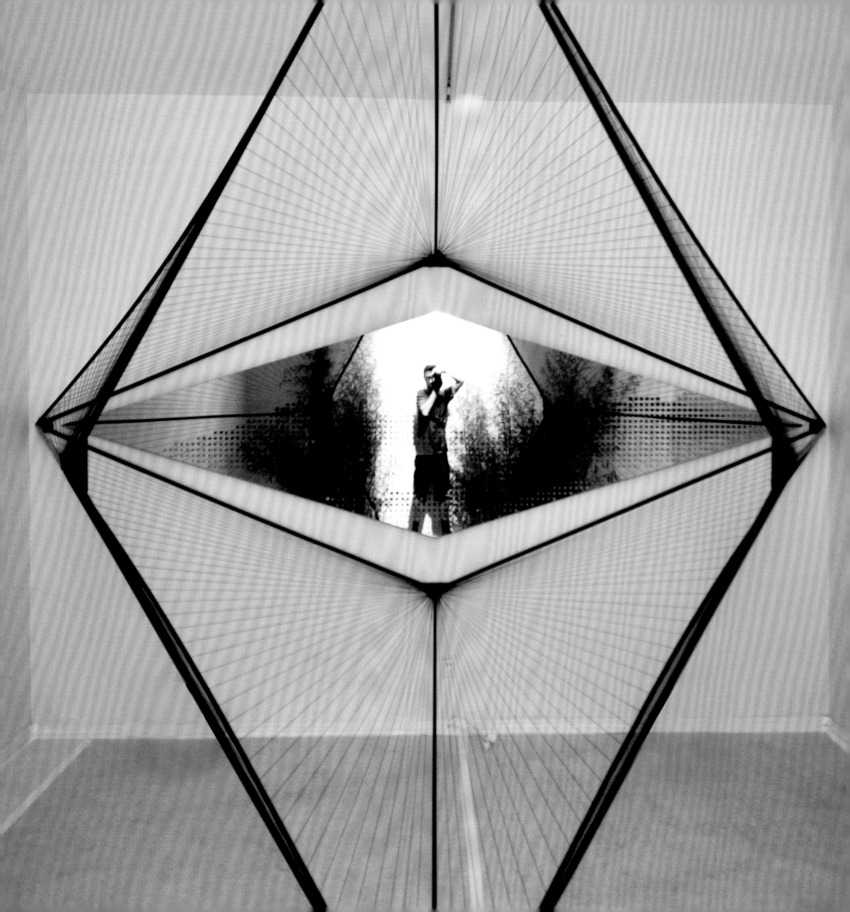

赵劲松
Zhao Jinsong

天津大学建筑学院/非标准建筑工作室
School of Architecture, Tianjin University/Non-standard Architecture Studio

视点与观点

这个世界本就是视点决定观点。无论是什么样的认知,本质上都是在编织一个解释世界的故事,差别源于立足点的不同。

在以"真实"为坐标的原始系统中所构建的认知,在以"虚拟"为坐标的系统里将会呈现出完全不同的样貌。

转换坐标,就是以新的方式重新构建与世界的连接方式,从而创造新的景观。同样的一件事,以不同的人为主角,就会呈现出完全不同的面貌。

这是一个共存的游戏:正交实体与斜交错觉的共存;室内现实与室外虚拟的共存;几何逻辑与艺术体验的共存;互动行为与空间装置的共存。

这是一个错觉的世界:你以为你所以为的就是你以为的吗?你以为你在屋里,其实你在窗外;你以为你在看山,其实你在山中。

你以为坐标是正交,但事实上是斜交;你以为装置是核心,其实观者才是作品。

Viewpoint and Point of View

In this world points of view are usually up to viewpoints. No matter what kind of cognition it is, it is essentially weaving a story that explains the world, and the difference stems from the difference in footing.

The cognition constructed in the original system with "reality" as the coordinate will take on a completely different appearance in the system with "virtuality" as the coordinate.

Converting coordinates is to re-construct the connection with the world in a new way, thereby generating a new landscape. The same thing, with different people as the protagonist, will show a completely different outlook.

This is a game of coexistence: the coexistence of orthogonal entities and oblique illusions; the coexistence of indoor reality and outdoor virtuality; the coexistence of geometric logic and artistic experience; the coexistence of interactive behavior and spatial installation.

This is a world of illusion: do you think what you think is what you think? You think you are in the house, but you are outside the window; you think you are looking at the mountains, but you are in the mountains.

You think that the coordinates are orthogonal, but oblique crossing is the real thing; you think the installation is the core, but the viewer is the work.

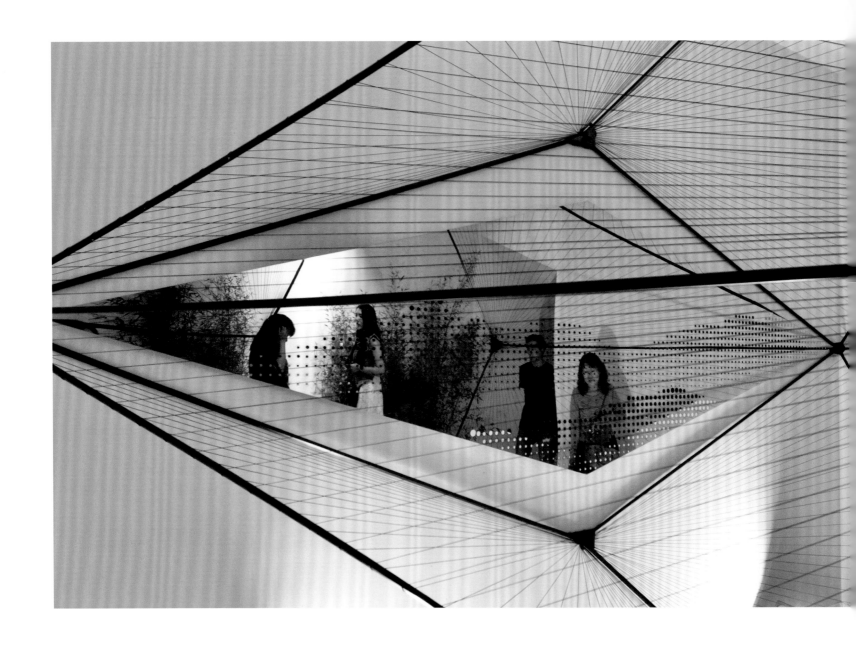

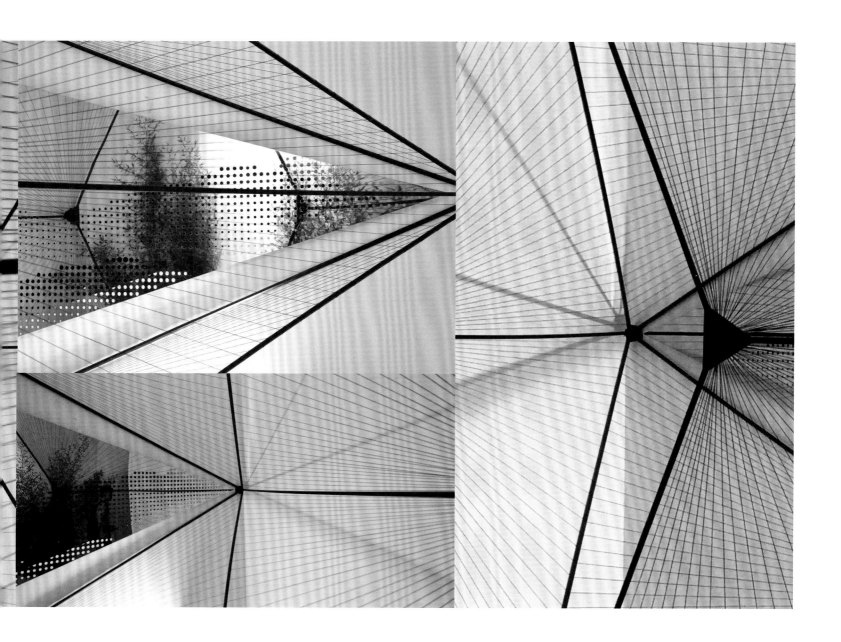

张大昕
Zhang Daxin

天津大学建筑设计规划研究总院有限公司
Tianjin University Research Institute of Architectural Design and Urban Planning Co., Ltd.

轮回之境

本次设计周的主题为"当代中的非当代",设计作品——"轮回之境"从两个方面来体现和诠释这个主题。一方面,从外观上来看,镜面中映出天光云影、亭台楼阁、太极古径……时而清晰,时而斑驳,置身其中,仿佛跟随这些历史片段回到了过去,从当代穿越到从前,去找寻古老的印记。另一方面,从设计理念来说,东南西北空间轮回,春夏秋冬四季轮回,生老病死命运轮回,年月更替兴衰轮回,没有起点,也没有终点,说不清从哪里开始,亦说不清在哪里结束。没有所谓的正反,只有走不完的轮回。历史与现在,现在与未来,都是一场场轮回。

装置由两组2m高的半圆形镜面不锈钢板组成,每组不锈钢板中间由四根钢龙骨支撑,钢板按照一定的曲率形成弧面,在两组钢板内侧人视点的位置上粘贴镜面贴膜。通过不锈钢板和镜面贴膜两种不同材质的穿插,周遭环境被反射成影影绰绰的景象,使整个装置充满了情境感和穿越感。在两组半圆形不锈钢板中央的地面上,放置一个正圆形的"太极"图案,"轮回"的概念进一步升华,表达了对宇宙、生命、物质、运动的思考与求索。

Reincarnation

The theme of this design week is "Non-contemporaneity in Contemporaneity". The design work—"Reincarnation" embodies and interprets this theme from two aspects. On the one hand, in terms of appearance, the mirror reflects the sky, clouds, pavilions, Tai Chi ancient path…Sometimes they are clear, sometimes they are blurred, as if following these historical fragments back to the past, traveling from the present to the old days, to find the ancient mark. On the other hand, space, seasons, fate, rise and fall are all reincarnations from the perspective of design concept. There is no start and no end. We could not tell where to start, and also can't tell where is the end. There is no so-called positiveness and negativeness, only endless reincarnation. The history and the present, the present and the future are all reincarnations.

The device consists of two groups of two meter high semi-circular mirror stainless steel plates. The middle of each group of stainless steel plates is supported by four steel keels. The steel plates form an arc according to a certain curvature. The mirror film is pasted on the position of human view point inside the two groups of steel plates. Through the interpenetration of stainless steel plate and mirror film, the surrounding environment is reflected into a vivid scene, which makes the whole device full of senses of situation and traveling. On the ground under the center of the two groups of semicircular stainless steel plates, a round "Tai Chi" pattern is placed to sublimate the concept of "reincarnation" and express the thinking and search for the universe, life, material and movement.

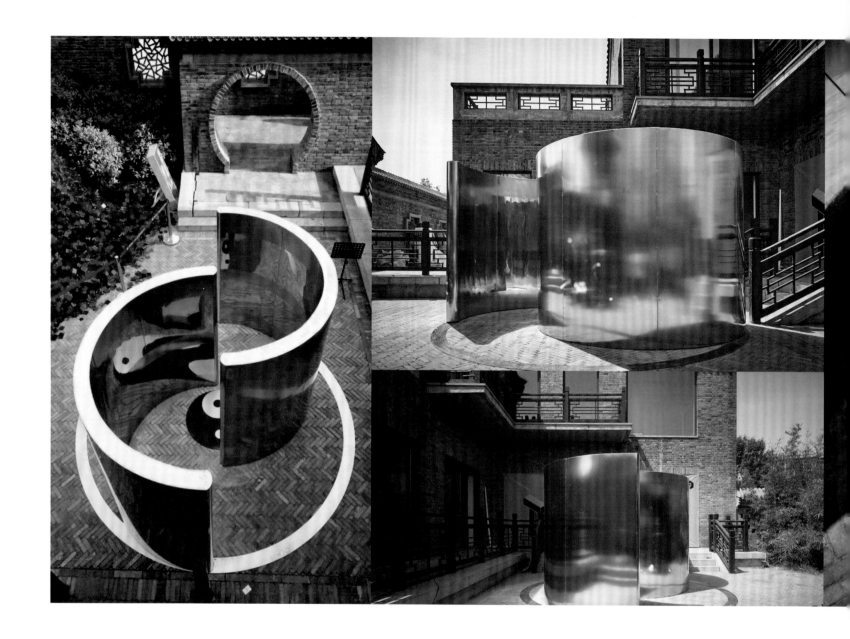
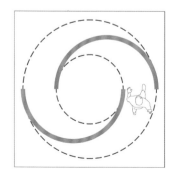
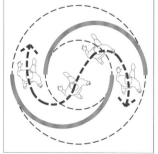

张曙辉
Zhang Shuhui

北京八作建筑设计事务所有限公司
Beijing Bazuo Architecture Office, Ltd.

玛尼堆中

一块块玛尼石堆叠的是千年岁月的沉淀和藏族群众永生的信仰,各层玛尼石之间往往有百年甚至千年的时间跨度。

"玛尼堆中"装置的设计理念来自石头的缝隙,这里纵贯千年。在利用镜子打造的无限空间中,石头脱离了重力,在其中可以感悟这种时间的跨度,感受生命的渺小。

In the Marnyi Heap

Marnyi stones are stacked with the sediment of thousands of years and the life-long belief of Tibetan. There are always time spans of centuries or even thousands of years between the layers of Marnyi stones.

The concept of "In the Marnyi Heap" comes from the gap between the stones, which runs through the millennium. In the infinite space created by the mirror, the stone is out of gravity, in which you can feel the span of time and feel the tininess of life.

2018年建筑展（作品展）
2018 Architecture Exhibition (Projects Exhibition)

2018

**2018年建筑展
（作品展）**
2018 Architecture Exhibition
(Projects Exhibition)

此间
——2018天津国际设计周建筑展综述
In Between
—Summary of Architecture Exhibition of
2018 Tianjin International Design Week

2018参展建筑师及作品
Participant Architects & Their Works

卞洪斌（天津大学建筑学院）
Bian Hongbin (School of Architecture, Tianjin University)

王宽（宽建筑工作室）
Wang Kuan (KUAN Architects)

任军（天津市天友建筑设计股份有限公司）
Ren Jun (Tianjin Tenio Architecture and Engineering Co., Ltd.)

庄子玉（BUZZ庄子玉工作室）
Zhuang Ziyu (BUZZ/Büro Ziyu Zhuang)

张华（天津大学建筑设计规划研究总院有限公司/张华工作室）
Zhang Hua (Tianjin University Research Institute of Architectural Design and Urban Planning Co., Ltd./Zhang Hua Studio)

张曙辉 王淼（北京八作建筑设计事务所有限公司）
Zhang Shuhui, Wang Miao (Beijing Bazuo Architecture Office, Ltd.)

卓强（天津市建筑设计研究院有限公司）
Zhuo Qiang (Tianjin Architecture Design Institute Co., Ltd.)

赵劲松（天津大学建筑学院/非标准建筑工作室）
Zhao Jinsong (School of Architecture, Tianjin University/Non-standard Architecture Studio)

鲍威（BWAO/鲍威建筑工作室）
Bao Wei (BWAO/Bao Wei Architecture Office)

此间——2018天津国际设计周建筑设计展综述
In Between
—Summary of Architecture Exhibition of 2018 Tianjin International Design Week

总策展人:宋昆
Chief Curator: Song Kun

策展人:张昕楠、胡子楠、赵伟、冯琳
Curators: Zhang Xinnan / Hu Zinan / Zhao Wei / Feng Lin

第五届天津国际设计周于2018年5月11—16日在天津市河北区北宁文化创意中心举行，主题为"东方×西方"（East×West），旨在展示东方文化与西方文化全新的、对等的交流与碰撞。作为本届设计周的重要单元，"此间（In Between）·2018天津国际设计周建筑展"由天津大学建筑学院宋昆教授团队进行策划和组织，团队成员包括张昕楠、胡子楠、赵伟、冯琳。

本次展览共邀请了9组京津两地的建筑师来展示他们近期的实践作品，包括卞洪斌（天津大学建筑学院）、王宽（宽建筑工作室）、任军（天津市天友建筑设计股份有限公司）、庄子玉（BUZZ庄子玉工作室）、张华（天津大学建筑设计规划研究总院有限公司/张华工作室）、张曙辉/王淼（北京八作建筑设计事务所有限公司）、卓强（天津市建筑设计研究院有限公司）、赵劲松（天津大学建筑学院/非标准建筑工作室）、鲍威（BWAO/鲍威建筑工作室）。其所在单位涵盖了国营设计院、民营企业、独立事务所，基本体现了不同类型建筑师在面对复杂城乡问题时所进行的敏锐思考以及应对之道。

当下中国建筑设计领域呈现一种开放的、多元的状态，本次展览的建筑作品从"宏大的奇观塑造"到"微观的空间更新"，从"传统延续"到"西方介入"，同一时期不同类型建筑师的作品成为社会发展的活态标本，"此间"的主题也是通过展现此时此地所发生的真实设计状态来对本届设计周的主题"东方×西方"进行回应。

The Fifth Tianjin International Design Week was held in Beining Cultural and Creative Center, Hebei District, Tianjin from 11th to 16th of May, 2018. Themed on "East×West", the event aimed to demonstrate brand-new exchanges and collisions on an equal footing between the Eastern and the Western cultures. "In Between · Architecture Exhibition of 2018 Tianjin International Design Week", as an essential unit of the design week, was planned and organized by Prof. Song Kun from the School of Architecture, Tianjin University and his team members included Zhang Xinnan, Hu Zinan, Zhao Wei and Feng Lin.

A total of nine groups of architects from Beijing and Tianjin were invited to display their recent design works in the Architecture Exhibition. They were Bian Hongbin (School of Architecture, Tianjin University), Wang Kuan (KUAN Architects), Ren Jun (Tianjin Tenio Architecture and Engineering Co., Ltd.), Zhuang Ziyu (BUZZ/Büro Ziyu Zhuang), Zhang Hua (Tianjin University Research Institute of Architectural Design and Urban Planning Co., Ltd./Zhang Hua Studio), Zhang Shuhui/Wang Miao (Beijing Bazuo Architecture Office, Ltd.), Zhuo Qiang (Tianjin Architecture Design Institute Co., Ltd.), Zhao Jinsong (School of Architecture, Tianjin University/Non-standard Architecture Studio), and Bao Wei (BWAO/Bao Wei Architecture Office). These architects came from diversified fields covering state-owned design institutes, private enterprises, independent firms and universities, thus basically reflecting the keen meditation and response of different types of architects in face of complex urban and rural problems.

The field of Chinese architectural design nowadays is characterized by an openness and diversity. The architectural works on display in this exhibition varied from "macro spectacle creation" to "micro space renovation", and from "continuation of tradition" to "Western involvement". The design works of different types of architects in the same period became the living incarnations of social development. And the theme "In Between" was therefore a response to the theme of the design week—"East×West" by showcasing the real design state that happens here and now.

2018   参展建筑师及作品
Participant Architects & Their Works

卞洪滨
Bian Hongbin

天津大学建筑学院
School of Architecture, Tianjin University

巴拿马佩里科岛阿马多尔邮轮码头客运中心

阿马多尔邮轮码头位于巴拿马佩里科岛的东北侧，航站楼位于佩里科岛的山丘和大海之间。

航站楼入口轴线与场地入口道路呈45°夹角，正对佩里科岛的山丘。航站楼呈U形布置，围绕中间的两层通高大厅缓缓地从地面升起，并盘旋上升。大厅内六组V字形的柱子撑起三角形主题的屋顶。透过大厅内三角形天窗能看到佩里科岛的山丘，同时面向大海的阳台为欣赏海景提供了良好的视点。航站楼以低矮的高度和流线的形体，谦逊地处于山与海之间，其造型充满了热带滨海的建筑特点。顶部表面覆盖光伏太阳能电池板，可以持续吸收太阳能。

Cruise Terminal of Amador, Isla Perico, Panamá

Cruise Terminal of Amador is located at the north-eastern side of Isla Perico, Panama and the terminal building is located between the hills of Isla Perico and the sea.

The entrance axis of the terminal is at a 45° angle to the site entrance road, facing the hills of Isla Perico. It is arranged in a U-shape, slowly rising from the ground, hovering around the middle of the two-storey-high hall. And the triangle-themed roof is supported by six groups of columns in V-shape. The hills of Isla Perico are visible through triangular skylights in the hall; at the same time, the balcony facing toward the sea provides a good view point to appreciate the seascape. The terminal, with low height and streamlined form, stands modestly between the hills and the sea, whose architectural shape is full of the characteristics of tropical coastal architecture.The top surface is covered with photovoltaic solar panels so as to absorb the solar energy continuously.

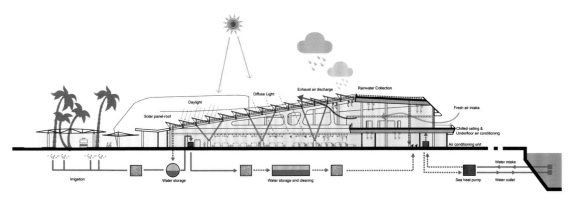

王宽
Wang Kuan

宽建筑工作室
KUAN Architects

青岛凤凰之声大剧院

薛家岛，古称"凤凰岛"，位于胶州湾西海岸青岛经济开发区黄岛区内，与团岛隔海相望，是国家级旅游度假区。相传古时候有一只金凤凰飞赴天庭参加百鸟盛会，飞抵碧波万顷、渔歌荡漾的胶州湾时，为此地美丽的海岛风光和淳朴的渔家民风所陶醉，乐不思飞，于是振翅落入胶州湾南侧，遂成今日薛家岛。因而凤凰自然而然地成为设计的出发点。

凤凰，亦作"凤皇"，是古代传说中的百鸟之王。雄的为"凤"，雌的为"凰"，总称凤凰，亦称为丹鸟、火鸟等，常用来象征祥瑞。凤凰齐飞，是吉祥和谐的象征，自古就是中国文化的重要元素。方案设计来源于凤凰这种生物，但不是其具体的形态，而是中国人心中对凤凰的情感。

仿生建筑的设计核心不只是形态，还有具备生物逻辑的结构设计。首先可以肯定的是，鸟类和哺乳动物生物体的骨骼关节系统以铰接体系为主，那么钢结构就是模拟其体系的最佳选择。设计的最佳形态是具有优越抗拉性能的线性钢结构构件，这和骨骼拥有非常相似的力学和形态特征。

"凤凰"的主体是两个矩形空间：多功能演艺大厅和千人歌剧厅。由于歌剧厅的配套空间更加复杂，且需要全封闭界面，因此将其隐藏在"凤凰"的后半身，在两侧单独设置了落地玻璃的出入口门厅。把"凤凰"的中心身体留给多功能演艺大厅，它通透、开敞，可进行充分的交流和互动。多功能演艺大厅平面被设计成一个圆形，在圆形的中央设置了一个中心舞台，四周环绕着观众座位。它是一个双向对称的大型无柱空间。最终的设计以四条巨型的钢拱撑起这个半球形的演艺大厅，由此"凤凰"的形式逻辑诞生了。四个钢拱落地的四对支脚为支点，生发出更多拱形，这些拱形伸出的跨度和高度各不相同，左右两侧对称生成"凤凰"的羽翼，并且一环一环向外扩散。前部向高远方向伸探到66m的空中，斜向悬挑近50m，却依然保持拱形逻辑不变。俯瞰建筑，其形态如花瓣一样往四个方向盛开。"凤凰"头部是一个神似于凤嘴的悬挑平台，在巨幅悬挑结构的端部进行第二次悬挑，将其设计为著名的海洋蹦极点，在这里人们可以肆无忌惮地观赏大海的翻滚和享受蓝天的拥抱，其端部被设计成一个空中餐厅。同时，"凤凰"的背部被设计为一个屋顶广场，通过180个台阶将广场与空中餐厅相连。登高望远的过程，仿佛是在云游四方。

一切鸟类，包括凤凰，最美轮美奂的是它们的身姿和羽毛。羽毛状的骨骼系统和放射状形态，成为这座仿生建筑的点睛之笔，同时，环绕建筑一周的、起伏变幻的、像风吹过其身体而掀起的律动一样的白色钢骨阵列，也对公共空间产生了遮阳效果。

The Sound of Phoenix Grand Theater

Xuejia Island, formerly known as "Phoenix Island" is located in Huangdao District, Qingdao Economic Development Zone, on the west coast of Jiaozhou Bay. It faces Tuandao across the sea. It is a national tourist resort. Legend has it that in ancient times, a golden phoenix flew to the heaven to participate in the Hundred Birds Festival. When it flew to Jiaozhou Bay, it was intoxicated by the beautiful island scenery and the unsophisticated fishermen's customs. So it landed on the south side of the Jiaozhou Bay, and became today's Xuejia Island. Therefore, the phoenix naturally became the starting point of the design.

The phoenix, is also known as the "Phoenix Emperor", the king of birds in ancient legends. The male and the female phoenix are collectively called phoenix. They are also known as Red Bird, Fire Bird, etc. It is often used to symbolize auspiciousness. Phoenixes flying together is a symbol of auspiciousness and harmony. It has been an important element of Chinese culture since ancient times. The schematic design started from such a creature as the phoenix, which is not a specific form but the emotion for the phoenix in the hearts of Chinese people.

The core design content of bionic architecture includes not only the form, but also the structural design with biological logic. First of all, it is certain that the bone and joint systems of birds and mammals are dominated by articulated systems, so the steel structure is the best choice to simulate their systems. The best form of steel structure is a linear member with superior tensile performance, which has very similar mechanics and morphological feature to bones.

The main body of the Phoenix is two rectangular spaces: a multifunctional performing arts hall and a thousand-person opera hall. Since the supporting space of the opera hall is more complicated and requires a fully enclosed interface, it is hidden in the back of the Phoenix, and separate entrance and exit halls with floor-to-ceiling glass are set up on both sides. The central body of the Phoenix is reserved for the multi-functional performing arts hall, which is transparent, open, and full communication and interaction can be carried out. The multi-functional performing arts hall is design to a circle, with a center stage set in its center, surrounded by audience seats. It is a large two-way symmetrical and column-free space. The final design is using four giant steel arches to prop up this hemispherical performing arts hall. It is precisely because of these four steel arches that the Phoenix's formal logic was born. The four pairs of feet on which the four steel arches lands as fulcrums to produce more arches. The protruding span and height of these arches are different. The left and right sides are symmetrical to form the wings of the Phoenix, and spread outward ring by ring. The front part protrudes into the air 66 meters high and cantilevered diagonally for nearly 50 meters, but the logic of the arch has remained unchanged. Seen from the aerial perspective, the architectural form is like petals blooming in four directions.The head of the Phoenix is a cantilevered platform that looks like the Phoenix's mouth. The second cantilever is performed at the end of the huge cantilever structure. Here is the famous ocean bungee spot, where you can enjoy and hug the sky and the sea as you wish. It is precisely because of this bungee spot that a restaurant in the sky is designed at the end. At the same time, the back of Phoenix is designed as a roof plaza, which is connected with the sky restaurant through 180 steps. The process of climbing high and looking in the distance seems to be wandering around.

For all birds, including phoenixes, the most beautiful is their appearance and feathers. The skeletal system and radial form of feathers have become the finishing touch of this bionic building. Simultaneously, an array of white steel bones that surround the building, undulate, and resemble the rhythm of the wind blowing over its body, have a shading effect on the public space.

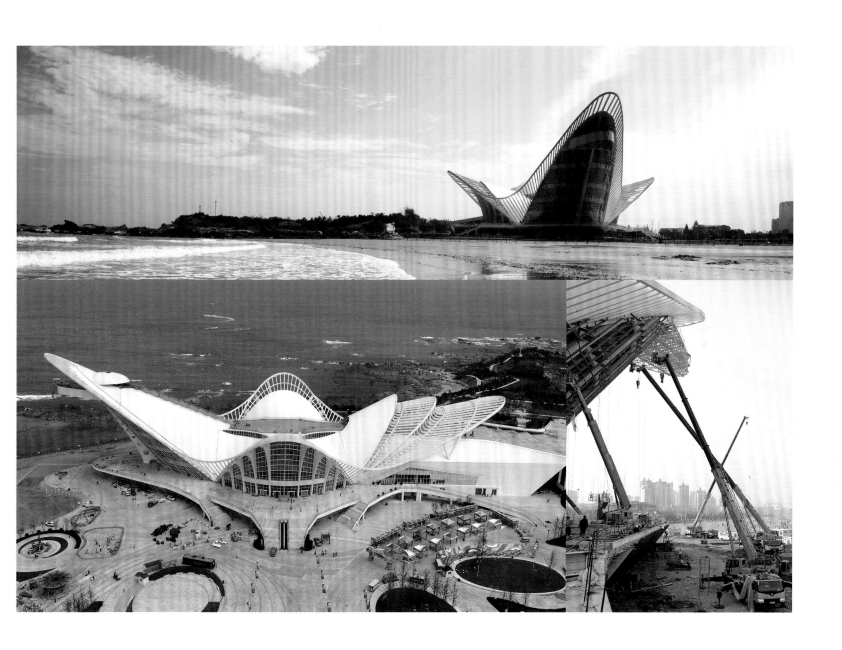

中新天津生态城健身馆

任军
Ren Jun

天津市天友建筑设计股份有限公司
Tianjin Tenio Architecture and Engineering Co., Ltd.

从绿色校园、绿色乡村到绿色改造

在这个气候变暖、空气污染的时代,绿色建筑是实现可持续发展的为数不多的选择之一。建筑师能做的还有很多……

绿色校园:生态教育——孩子们将在健康环境中接受潜移默化的生态教育。

绿色乡村:低碳乡居——创造零煤耗的低碳农宅居住环境。

绿色体育:健康建筑——健康与节能是绿色体育馆的双核。

绿色改造:超低能耗——绿色技术集成实现国际水准的超低能耗。

From Green School, Green Village to Green Transformation

Green building is one of the few choices to achieve sustainable development in this era of global warming and air pollution. There are a lot of things that architects can do...

Green campus: ecology education. Children will receive an imperceptible eco-education in a healthy environment.

Green village: low carbon village house. It is to achieve low carbon residential environment with zero coal consumption.

Green gym: healthy building. Health and energy conservation are the dual core of green gym.

Green transformation: low energy consumption. It is to integrate green technologies to achieve ultra-low energy consumption at an international level.

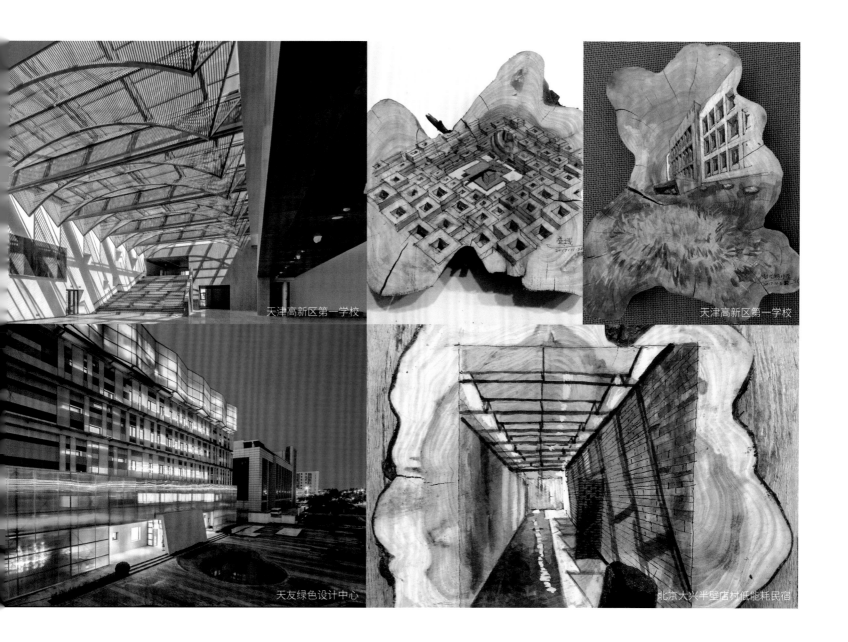

天津高新区第一学校　　天津高新区第一学校

天友绿色设计中心　　北京大兴半壁店村低能耗民宿

庄子玉
Zhuang Ziyu

BUZZ庄子玉工作室
BUZZ/Büro Ziyu Zhuang

谷仓博物馆

谷仓博物馆将会被建于挪威克里斯蒂安桑市。以保护、更新、添补为设计概念，一个历史悠久的谷仓被融入这座新的博物馆综合体中。新加入的体量赋予了这个高挑的谷仓别具一格的特色，河边的一个正方形"艺术管道"与水面相连接，塑造了富有吸引力的视觉效果。通过河边漫步区，可从由表演艺术中心与博物馆组成的文化中心过渡到更加平静的生活居住区。在陆地一侧，从谷仓延伸出来的长体量被用作新广场的核心，它横跨谷仓，作为新建筑，也是管道。在已经存在的谷仓建筑中挖出一个弧球体量，这样既为游客创造了一个庇护广场，又可成功地展示其美丽的内部空间。从那里，游客可以追随着粮谷的步伐，从建筑的顶层开始，从东边的新建筑迂回，逐渐走到底楼。与完全封闭的弧状混凝土谷仓形成鲜明对比，这座新建筑被透明开放的玻璃包裹。多媒体的建筑立面既可以呈现半透明形态，展现室内的投影轮廓，也可以展示从诺吉·坦根收藏品中精选的艺术展品，阐述了从"私人收藏"迈向"大众艺术"的艺术理念。

Kunstsilo Museum

Kunstsilo Museum will be built in Kristiansand, Norway. Following a concept born from the mix of preserving, modifying and adding, a historic grain silo is integrated into the new museum complex. Newly added volumes give a distinctive character to the long and thin silo building. A rectangular "art tube" is featured on the waterside, pushing itself into the water as a visual attraction. The cultural center consisting of the performance art center and the museum transits to a calmer residential neighborhood through the waterfront promenade. On the landside a long extension of the silo serves as the core for the new plaza, which spans the silo, as both a new building and a canal. The existing silo is transformed through cutting a curved shape out of it, achieving both goals of creating a sheltered forum for visitors and revealing the inner beauty of the building. The visitors can follow the former way of the grain: the exhibition loop starts on the top of the building and gradually makes its way downwards, via detour of the new building on the eastern side. In stark contrast to the closed, curved concrete silo, this new building is wrapped in transparent and open industrial glass. Its skin acts as a media facade, either appearing translucent to display the shadow of its interior, or displaying selected artworks from the Norge Tangen Collection, elaborating an artistic idea from "private collection" to "public art".

南京丰盛五季凯悦臻选酒店

南京丰盛五季凯悦臻选酒店呈现自东向西打开的半围合布局;针对地势起伏较大的地形条件,利用现代造园手法表现传统园林韵味。通过随形就势,柔化形态,打造三层曲面环廊,以水、草木布局为中心构建下沉景观庭院。建筑沿中心庭院处做了大量的底层架空和屋顶覆土,将自然光线与景观尽可能地引入建筑空间当中。以回廊、折廊、曲廊等多种形式划分庭院空间,产生一步一景、移步景异的观感和达到多样统一的空间效果。酒店东、南、北侧的建筑通过单、双面空廊与地面连接,既将分散的园林小景联系在一起,同时又组织和引导行进的路径。透过酒店门窗可看到白墙黄瓦、古朴肃穆的天隆寺;浡泥国王墓四周郁郁葱葱的山林四时之景亦被框入景中,从视觉层面巧妙地将古与今、建筑与自然相互连接。自由开朗的空间布局、立面引入的格栅体系、水木置石的点缀,加深了视觉的纵深感,流动空间无我之境的意境自然产生。穿过上下起伏的下沉庭院,由西而东,循着流水与置石,另辟一处蹊径,来到水院禅堂。水院禅堂采用清水混凝土的结构模式,南北客房采用标准框架结构,东侧采用纯钢结构桁架体系悬挑,形成大跨度空间,营造一种仪式感。同时将结构反梁植入空中庭院,强化建筑的漂浮感,使景观从地面抬升到高位空间,从视觉上丰富建筑与自然的层次感,呼应人与自然的联系。大量的覆土种植、园林景观不断软化建筑界面,消解建筑的体量。通过抬升、架空,建筑体块漂浮在酒店上空,产生了更多的公共空间,增加了人的流动性,加强了景观的流动性以及空间的流动感。

南京丰盛五季凯悦甄选酒店

Grand WUJI Hotel, the Unbound Collection by Hyatt, Nanjing

Grand WUJI Hotel, the Unbound Collection by Hyatt, Nanjing presents a semi-enclosed layout that opens from east to west. Aiming at the situation of fluctuations in terrain, modern gardening methods are used to express the charm of traditional gardens. By softening the shape with the current terrain, a three-layer curved circular corridor is created, and a sunken landscape courtyard is constructed which is centered on the layout of water and vegetation. The building has made a large amount of ground floor on stilts and roof covering along the central courtyard to bring natural light and landscape into the building space as much as possible. The courtyard space is divided into various forms such as cloisters, folding corridors and curved corridors, to achieve a different perception of scenery with a step in a scene and a unified space effect. The east, south and north sides of the hotel are connected by single and double-sided colonnades, which simultaneously connect the scattered garden scenery and organize the path of walking. The doors and windows enframe the scenery of the solemn Tianlong Temple with white walls and yellow tiles, and the lush mountains and forests surrounding the tomb of King Boni. From the visual aspect, the past and the present, the architecture and the nature are cleverly combined. Wide spatial layout, the grid system brought in the facade, and the embellishment of water, wood and stone, deepen the sense of vision and the artistic conception of the flowing space is naturally produced. Passing through the undulating sunken courtyard, from west to east, and following the flowing water and the stone, a new path is created to the meditation hall. This meditation hall adopted the structural model of fair-faced concrete. Guest rooms on the north and the south used a standard structure, and the pure steel truss system cantilevered on the east forms a large span space, which creates a sense of ceremony. At the same time, the reverse beams were implanted in the sky courtyard, strengthening the floating sense of the building, so that the landscape rises from the ground to higher space, visually enriching the sense of hierarchy between the building and nature, echoing with the connection between man and nature. A large number of soil-covered plantings and garden landscapes constantly soften the interface and dissolve the volume of the building. Through lifting and overhead methods, the building block floats above the hotel, creates more public space, increases the mobility of people, and strengthens the mobility of the landscape and the mobility sense of space.

金山岭书院

金山岭书院是对长城——这个最具中国文化意象的遗产的一种延续，表现为书院与长城之间产生的时空层面的对话，暗示一种文脉相承的隐喻关系。在具体操作形式层面，项目本身具有非常强的轴线关系和当代调性，同时它整合了古典园林散点建筑与场地的关系，形成了具有传统东方仪式感的品质空间。整个项目从平面关系上被分成三段：第一段是中国传统游廊空间的变异，其本身接近于游园，在自由愉悦游走的状态下产生世俗化的空间体验。中间设有宽敞别致的中央庭院，沿着游廊形成一系列步移景变的景观效果，形成连续的、具有画面感的空间连续体。在穿过墙院体系的过程中，光影在变化，空间的纵深也在变化，回应了东方院落文化的空间体验。在整个体系里，人们在一种开敞、压抑，再开敞、再压抑，然后再度开敞氛围中，在室内、室外、半室内的空间内游走。这一系列空间结合游走的顺序和体验式的探索逐层递进，在游览者眼前慢慢展开了一幅散点透视、时空交织的长卷。这幅长卷能够在叙事性层面上呈现给受众一种渐进式的体验。

金山岭书院

Jinshanling Academy

Jinshanling Academy is an continuation of the Great Wall, a cultura relic which has the typical Chinese cultural imagery. The dialogue between the academy and the Great Wall at the space-time level implies a metaphorical relationship of context. As for the specific operational form, the project itself has a strong axis relationship and contemporary tonality. At the same time, it integrates the relation of classical garden scattered architecture and the site to form a quality space with traditional oriental ritual feeling. The whole project is divided into three sections from the plane relation: the first section is the variation of the space of the Chinese traditional verandah, which is similar to visiting a garden, and produces a secularized spatial experience under the condition of free and pleasing wandering. There is a spacious and unique central courtyard in the middle, and a series of step-shifting landscape effects along the verandah appear, forming a continuous and pictorial spatial continuum. In the process of passing through the wall and courtyard system, the light and shadow is changing, the depth of the space is also changing, which is a response to the space experience of Oriental courtyard culture. Throughout the system, people move from openness to repression, then from openness to repression, finally to openness once more, in indoor, outdoor, and semi-indoor space. This series of space combines the order of wandering and the exploration of experience in layers. It slowly unfolds to the visitor a long scroll of scattered perspective, interwoven with space and time. This long scroll is intended to present to the audience a progressive experience at the narrative level.

铜陵山居

铜陵山居原本是一幢融合徽州与沿江风格的普通民居,地处皖南一个偏远山村。原有建筑处于全村最高的山顶上,占地较小,且极为残破,已逾10年未有人居住,故其四周皆为杂草和灌木覆盖;其东西向三跨,南北向一跨,原有屋面和墙体损毁严重。故而从平面布局开始,本设计考虑西向增加一跨直抵一侧岩石山体,南北向增加一跨以形成较宽阔的起居室空间;并在平面上部分引入曲线,一个异化的虚体从原有投影轮廓中抽离出来。由于原有层高的限制和原有屋面完全破损不能再用,本设计选择将原有建筑加高至两层,并结合平面上抽离出的双生虚形在空间上放样拉伸,将前面的部分脊线压低,形成前后错落的空间连续曲面。传统的折面屋面和旁边抽离出来的流线型空间融合成一体,暗合了中国文化"道生一,一生二"的宇宙观。从鸟瞰图来看,整个屋面以青瓦覆盖,形成了独特的形态,与现有古村落既融合又出挑,并首次将室内的空间特征从外部表述出来。

如此形成了一个东西向的四跨空间。从东起,一跨形成地块南向的作为起居室的前厅空间,一跨与横向展开的部分残墙形成庭院空间及二层的玻璃观景平台,一跨与西侧的原建筑墙体形成卧室空间,增加一跨向西将山体和营造的部分景观空间纳入半开放的檐下空间,由此在立面上构成一幅传统语境下的当代拼接画。

Tongling Mountain House

Tongling Mountain House is located in a remote village in the south of Anhui Province, which was originally a normal folk house blending the style of Huizhou and the style along the river. Nested on the top of the mountain in the village, this small house was in an extremely bad condition covered by weeds and bushes and uninhabited for over ten years. It had three bays in the east-west direction and one bay in the south-north direction. Its original roofing and wall were badly damaged. Starting from the plan layout, the design considered to add one bay to the west to reach the rocky mountain on one side, and another bay in the south-north direction to form a wider living space. The design introduced partly curves in the plane to create an imaginary space detached from the original projection. Because of the limitation of original height and badly damaged roof, the design added a new floor to the original building. Together with the imaginary space detached from the original projection, an open spatial view was created. The disign lowered part of ridge in the front to form an uneven continuous curved silhouette. Traditional folded roof and the detached streamlined space were blended together, coinciding with the cosmology of Chinese culture: "The Dao (Way) produces One (world), One produces Two (Yin-Yang)." The whole roof was covered by grey tiles in a unique shape from a bird's eye view, which was not only outstanding but also a convergence with existing ancient village. This was the first time to echo with the features of indoor space in outdoor areas.

The four bays forms a space in the east-west direction. From the east, one bay is used as the front hall next to the living room in the south. One bay and the transverse part of the residual wall form a courtyard as well as the glass platform on the second floor. One bay combined with the original wall in the west forms a space for the bedroom. The added bay embraces the mountain view to the west and takes part of the built landscape into its semi-open virtual space under the eaves. As a result, the elevation presents a modern collage picture under traditional context.

铜陵山居

小米醋博物馆

张华
Zhang Hua

天津大学建筑设计规划研究总院有限公司/张华工作室
Tianjin University Research Institute of Architectural Design and Urban Planning Co., Ltd./Zhang Hua Studio

张华工作室坚持设计创新，走中国建筑原创道路，具有世界级的设计水平。其建筑理论源于中国道教的自然阴阳观，以及中国独特的诗词、书画、赏石等山水文化艺术。现代数学的分形几何理论，拓扑、物理及生物等科学是其建筑设计理论的另一来源。他提出了"流形""分形"和"拼形"三个理论概念，首创分维建筑，兼具参数化设计和3D打印技术，初步建立了个人风格和流派。

工作室代表作品有天津蓟州国家地质博物馆、陕西宝鸡青铜器博物院、柳州奇石馆、于庆成美术馆及小米醋博物馆等。

Zhang Hua Studio adheres to design innovation, takes a path of original Chinese architecture, and is at the world-class design level. Zhang Hua's architectural theory comes from the natural Yin-Yang view of Chinese Taoism, as well as the unique Chinese poetry, calligraphy and painting, stone appreciation and other landscape culture and art. Fractal geometry theory of modern mathematics, topology, physics and biology are another source of his architectural design theory. He put forward three theoretical concepts: manifold, fractal and collage, and created fractal architecture with parametric design and 3D printing technology, initially establishing a personal style and school. The design works have won dozens of awards at home and abroad.

The representative works of the studio are Tianjin Jizhou National Geological Museum, Shaanxi Baoji Bronze Ware Museum, Liuzhou Stone Museum, Yu Qingcheng Art Museum, Millet Vinegar Museum and so on.

小米醋博物馆

柳州奇石馆　　天津武清永定河故道国家湿地宣教中心

丁青县藏医院

张曙辉
王淼
Zhang Shuhui
Wang Miao

北京八作建筑设计事务所有限公司
Beijing Bazuo Architecture Office, Ltd.

丁青县藏医院

台地的巧妙处理：10m断坎为基地不利因素，该设计利用屋顶花园拓展断坎以上的活动空间，利用建筑减少护坡面积，形成独有空间和景观。

1. 传统的层间布局：西藏传统建筑分为三层，上—神—治愈，中—人—住院，下—牲畜—支持，本次设计呼应西藏传统文化。

2. 藏医的现代化：传统藏医仅有治疗空间，该设计采用现代工艺优化藏医治疗，实现"患者—治疗—医护"的模式。

拉萨市中心医院

1. 零能耗：针对西藏脆弱的生态环境，地下室采用自然通风的"零能耗"模式。

2. 藏式院落：现代的医疗系统规划，呈现的却是传统雪城的回字形院落。

3. 生长：医疗主街设计为可南北生长的模式，为未来发展留足空间。

4. 标准化：8.1 m的柱网贯穿景观、建筑及室内。

Tibetan Hospital in Dingqing County

Ingenious use of the terrace: The 10-meter height difference of the terrace is a disadvantage within the site. In this project, the roof garden is used to expand the activity space above the terrace, and the building used to reduce the retaining slope area, which forms a unique space and landscape.

Traditional floor layout: Traditional Tibetan building is divided into three levels: upper -god -healing, middle - human - residence , and lower - livestock - support. This project respects and echoes with this traditional Tibetan culture.

Modernization of Tibetan Hospitals: In traditional Tibetan hospitals, there is only treatment space. In this project, modern technology is used to optimize the traditional Tibetan hospital, which realizes the system of "patient -treatment - medical staff".

The Central Hospital of Lhasa

Zero Energy Consumption: In view of the fragile ecological environment in Tibet, the "Zero Energy Consumption" mode of natural ventilation is adopted in the basement space.

Traditional Tibetan courtyard layout: The modern medical system planning presents a traditional Tibetan courtyard layout of "Hui"(回).

Growth: The main medical street is designed as a north-south expansion system, which leaves enough space for future development.

Standardization: 8.1m modules are used in landscape, architecture and interior design to achieve standardization.

丁青县藏医院

拉萨市中心医院

卓强
Zhuo Qiang

天津市建筑设计研究院有限公司
Tianjin Architecture Design Institute Co., Ltd.

迦陵学舍
迦陵学舍是南开大学为国学大师叶嘉莹先生建造的集科研、教学、办公于一身的现代书院。中国国画被誉为"凝固在二维空间里的人生",散点的布局将一个个片段组合成整体,透射出特有的文人情怀。方案用融、障、透、框的手法在不同节点形成一连串精心设定的画面,再用流线将这些画面串接,希望以国画传统的手法向国学大师致敬。

宁夏美术馆
21世纪的美术馆作为艺术载体,本身应具有深厚的文化艺术气质。同时,如何体现当地的地域特征也是设计着重思考的方面。该案试图通过写意的方式,用简洁的建筑语汇赋予宁夏美术馆一种独特的气质,同时给观者一个开放的平台和自由的想象空间。

天津城建大学图书馆
对于现代大学生活来说,图书馆不仅是一个求知的圣地,更是一个社交的场所,在这里学生既能经历"书山有路勤为径"的艰辛,也能感受"最是人间堪乐处"的美好。这种二元相融相生的关系恰似中国传统木构的工艺精髓——榫卯。新图书馆正是以"书山径、榫卯间"为概念,为该所大学打上了深深的城建烙印。

Jialing House
Jialing House is a modern academy integrating scientific research, teaching with working, which was built for Ye Jiaying, a master of Chinese culture in Nankai University. Chinese traditional painting is known as the life solidified in the two-dimensional space. The splattering layout puts the pieces into a whole, casting specific humanistic feelings. By means of techniques such as melting, barrier, penetrating and framing in this case, the architecture at different nodes form a series of delicate images, which are connected by streamline. We want to salute to the master of Chinese culture with traditional techniques of traditional Chinese painting.

Ningxia Art Museum
As the carrier of art in 21 Century, the art museum itself should have a profound cultural and artistic temperament. At the same time, how to reflect the regional characteristics is also the focus of the design. The case attempts to give the Ningxia Art Museum a unique temperament with concise architectural vocabulary, and to give the viewer an open platform and a space for free imagination.

Tianjin Chengjian University Library
For modern university life, library is not only a holy land for knowledge, but also a place for socializing. In the library, people can experience the hardship of diligence which is the only way to get to the mountain of knowledge. At the same time, it can offer the best place to live. This kind of dual relationship is similar to the essence of Chinese traditional wooden craft, mortise and tenon. The new library is designed with the concept of 'the book trail, the mortise and tenon joint', which brings deep brand of the spirit of ChengJian University.

南开大学迦陵学舍

北京世界园艺博览会生活体验馆

赵劲松
Zhao Jinsong

天津大学建筑学院/非标准建筑工作室
School of Architecture, Tianjin University/Non-standard Architecture Studio

北京世界园艺博览会生活体验馆

项目为北京世界园艺博览会生活体验馆,位于北京市延庆区西南部,设计概念源于对植物纤维的演绎。本方案用纤维构成一个似山非山、似房非房的朦胧界面。它是一个介于室内与室外、建筑与景观之间的装置。上部,既似远山云影,又似屋顶曲面;中部,既似树木成林,又似空间柱列;下部,既似广袤大地,又似层层台基。场地在这个朦胧装置的覆盖下形成一个奇妙的场所。从外往里看,叠映的山影与自然山水交相辉映;从里往外看,如同透过朦胧的雨雾审视山水。人在其中既可以神游于山水之外,又可以置身于山水之中。

合一与共在

在光与影之间,影与光同在。

在内与外之间,外与内同在。

在无与有之间,有与无同在。

在实与空之间,空与实同在。

Life Experience Hall Of Beijing World Horticultural Expo

This project is the life experience hall of Beijing World Horticultural Expo, located in the southwest of Yanqing District, Beijing. The design concept is derived from the deduction of plant fiber. In this scheme, fiber is used to form a hazy interface that looks like a mountain or a house. It is a device between indoor and outdoor, and between architecture and landscape. The upper part is like the cloud shadow of distant mountains or the curved surface of the roof. The middle part is like a forest of trees or a spatial colonnade. The lower part is like a vast land or layers of platform foundation. The site forms a wonderful place under the coverage of this hazy device. Seen from the outside to the inside, the superimposed mountain shadow and natural landscape complement each other; from the inside to the outside, it is like looking at the landscape through the hazy rain and fog. People can not only wander outside the mountains-and-waters, but also stay in the mountains-and-waters.

Unity and Coexistence

Between light and shadow, shadow and light exist together.

Between inside and outside, outside and inside exist together.

Between nothing and everything, everything and nothing exist together.

Between solid and void, solid and void exist together.

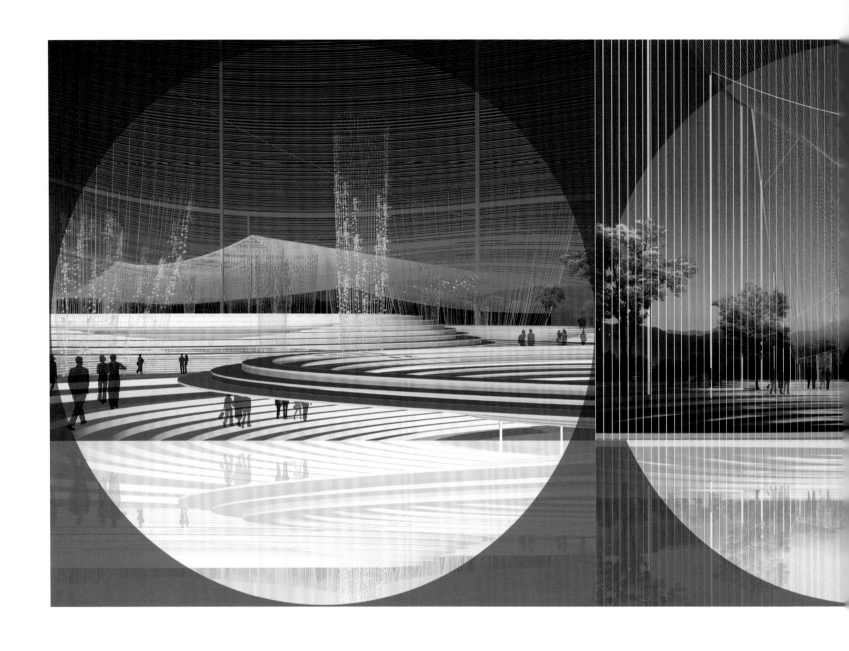

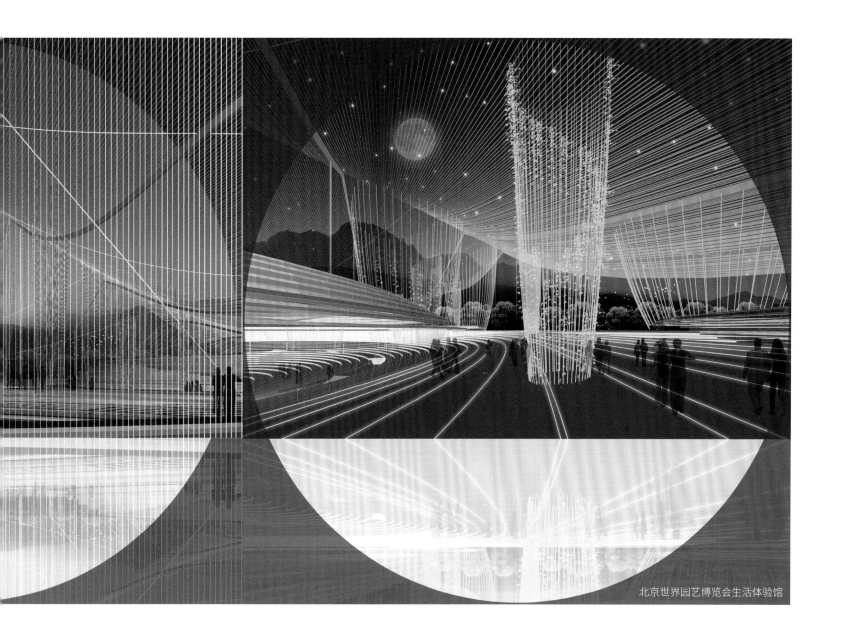

北京世界园艺博览会生活体验馆

烂缦胡同66号院

鲍威
Bao Wei

BWAO/鲍威建筑工作室
BWAO/Bao Wei Architecture Office

烂漫胡同66号院

烂漫胡同66号院是典型的大杂院，其改造的范围为南侧两间正房和北侧一间自建房。正房为典型的木构屋顶，无南向采光。自建房年久失修，在此次改造中重建。

五个大小与功能不同的体块被插入现有场地，与正房原始结构构成新的居住空间。东房被改造为工作室，对应的插入体块为东侧的卫生间与北侧的工作台；西房被改造为居住空间，对应的插入体块为北侧的起居室；西房北侧的厨房体块与起居体块屋顶相连；北侧的自建房体块被改造为餐厅，上方3m×1.1m的通长天窗仅用于采光，而且将枣树、屋脊与天空纳入景框构图之中，使这个空间成为家庭活动的精神堡垒。

No.66 Lanman Hutong

No.66 Lanman Hutong is a typical courtyard shared by many households. The scope of this renovation includes two main rooms on the south side and one self-built structure on the north side.The main room has a typical wooden roof with no southern lighting.The self-built house was in disrepair, so it was rebuilt in this renovation.

Five blocks in different sizes and with different functions are introduced to the site, forming a new living space with the existing structures. East room is transformed into a studio by inserting a bathroom block to the east and a working station block to the north. West room is converted to a residential space by attaching a living block to the north. A kitchenette block to the north is connected with the living block via a roof canopy. The self-built structure is reconstructed and turns into a dining space. A clerestory window with a dimension of 3m×1.1m not only brings light into the space, but also takes the jujube tree, the roof ridge and the sky into the picture frame composition, making the dining space a pivotal point for family activities.

烂缦胡同66号院

望京SOHO真传泰拳健身馆

望京SOHO真传泰拳健身馆

望京SOHO以强势的建筑姿态干预并主导着望京地段的城市空间。此次室内设计的场地东、南、北三面面向城市打开。室内如何与建筑协同、与城市对话是本设计的出发点。设计将拳台布置在视觉的焦点，这与经典剧场平面的概念如出一辙：拳台即舞台。场地上的七根柱子被连续的曲面包裹，在满足功能需要的同时，形成了空间的连续性。拱的存在与城市界面形成了一个应答，将泰拳这种竞技运动象征性地带入角斗场的意象中去。

Wangjing SOHO Muay Thai Gymnasium

Wangjing SOHO dominates the urban space in Wangjing section with its strong architectural presence. The site for this interior design is open to the city on three side including east, south and north. The starting point of this design is how the interior works in collaboration with the building and to have a dialogue with the city. The boxing ring is placed at the focal point of the space, similar to a classic theater in plane: the ring is the stage. The seven columns on the site are wrapped up by a continuous curved surface, not only meeting the functional needs but also forming a spatial continuity. The existence of the arch echoes with the city interface, symbolically taking the competitive sports of Muay Thai into the imaginary of arena.

2019 年建筑展（作品展）

2019 Architecture Exhibition
(Projects Exhibition)

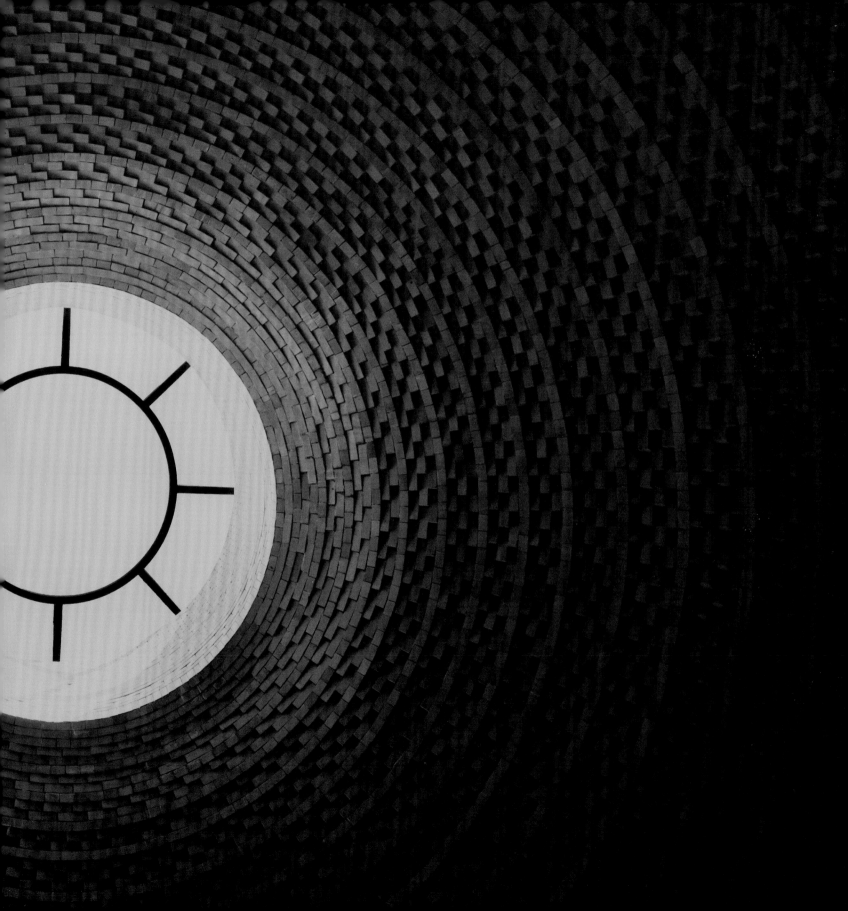

2019

2019年建筑展
（作品展）
2019 Architecture Exhibition
(Projects Exhibition)

汇流而下的城乡之辩
——2019天津国际设计周建筑展综述
The Confluence: the Debate between Urbanity and Rurality
— Summary of Architecture Exhibition of 2019 Tianjin International Design Week

2019参展建筑师及作品
Participant Architects & Their Works

张华（天津大学建筑设计规划研究总院有限公司/张华工作室）
Zhang Hua (Tianjin University Research Institute of Architectural Design and Urban Planning Co., Ltd./Zhang Hua Studio)

赵劲松（天津大学建筑学院/非标准建筑工作室）
Zhao Jinsong (School of Architecture, Tianjin University/Non-standard Architecture Studio)

狄韶华（第一实践建筑设计）
Di Shaohua (PRAXiS d'ARCHITECTURE)

任军（天津市天友建筑设计股份有限公司）
Ren Jun (Tianjin Tenio Architecture and Engineering Co., Ltd.)

郭海鞍（中国建筑设计研究院乡土创作研究中心）
Guo Hai'an (Rural Culture D-R-C, China Architecture Design & Research Group)

韩文强（中央美术学院建筑学院/建筑营设计工作室）
Han Wenqiang (School of Architecture, Central Academy of Fine Arts/ARCH STUDIO)

陈天泽 关英健 甄明扬（天津市建筑设计研究院有限公司）
Chen Tianze, Guan Yingjian, Zhen Mingyang (Tianjin Architecture Design Institute Co., Ltd.)

卜骁骏 张继元（时境建筑设计事务所）
Bu Xiaojun, Zhang Jiyuan (Atelier Alter Architects PLLC)

田恬（天津市城市规划设计研究总院有限公司）
Tian Tian (Tianjin Urban Planning & Design Institute Co., Ltd.)

张曙辉 王淼（北京八作建筑设计事务所有限公司）
Zhang Shuhui, Wang Miao (Beijing Bazuo Architecture Office, Ltd.)

王求安（北京安哲建筑设计有限公司）
Wang Qiu'an (Beijing ANT Architectural Design Co., Ltd.)

张东光 刘文娟（合木建筑工作室）
Zhang Dongguang, Liu Wenjuan (Atelier Heimat)

申江海（大观建筑设计）
Shen Jianghai (Daga Architects)

汇流而下的城乡之辩——2019天津国际设计周建筑展综述

The Confluence: the Debate between Urbanity and Rurality — Summary of Architecture Exhibition of 2019 Tianjin International Design Week

总策展人：宋昆
Chief Curator: Song Kun

策展人：黄元炤、张昕楠、胡子楠、赵伟、李德新
Curators: Huang Yuanzhao / Zhang Xinnan / Hu Zinan / Zhao Wei / Li Dexin

天津是环渤海地区的重点城市,拥有着丰富璀璨的建筑文化。历史建筑遗产是构成天津建筑文化的重要一环,如老城的广东会馆、租界的万国建筑,与新建筑共同汇聚、塑造了中西合璧、古今兼容的城市风貌,为天津的城市、建筑与空间环境发展提供了重要的财富与资源。这些建筑遗产在城市更新过程中扮演的角色也相当重要,成为天津申请设计之都的重要支撑。此外,在当代中国的建筑图景之下,无论城市、乡村,还是城乡之间,存量与增量是设计前要面对与思考的先决条件之一,因此建筑师会从此视角切入,用设计智慧解决项目在不同层面与角度的实际问题与需求。

2019天津国际设计周建筑展,由天津国际设计周建筑展总策划、天津大学建筑学院党委书记宋昆教授和策展人黄元炤、张昕楠、胡子楠、赵伟、李德新共同构思,用彰显天津建筑文化的"汇流"作为展览主题,汇聚京津地区的实践建筑师,共同思考存量与增量之于项目的变因,用作品展示与现场交流的方式展开对主题的讨论,并试图创建新的思维。策展团队共邀请了13组建筑师团队参展,他们分别来自设计院(张华、任军、郭海鞍、陈天泽、关英健、甄明扬、田恬)、事务所(狄韶华、卜骁骏、张继元、张曙辉、王淼、王求安、张东光、刘文娟、申江海)与高校(赵劲松、韩文强)。

本次展览以"汇流"为主题,且以存量与增量为背景支撑,策展团队更多地关注旧建筑的改造,而非新建筑,即在存量的基础上对项目做适度的增量,并关注改善、激活与新增等方面。不仅如此,策展团队还决定提出符合当下建筑图景发展的二元区块论,把改造分为城市改造与乡村改造两部分,进而布置了"城市改造"与"乡村改造"两大展示板块,其各自成形,形成两大支脉,共同汇流在建筑展之内。策展团队希望把建筑展塑造成一分为二、非此即彼的展示板块,形成一种暂时对立的"虚拟"状态。从概念表达到实践演示,加强对城乡二元真实性与现实性的辩证讨论,在交叉论证之下让展览主题的核心价值更为清晰。在两大展示板块确立后,参展建筑师根据他们的设计实践及其参与改造项目的比重,分别挑选了自己所感兴趣的板块。两个板块之间的对话体现出"城乡之辩",使建筑展的最终目的得以善意地呈现。

城市与乡村是两类人居环境。在中国,城市与乡村通过一系列建设展现各自的优势。城市经由改革开放后的现代化建设,用地扩张,高效建造,产生了许多城市建筑,建立了城市精神文明。同时,经由乡村振兴战略的实施,乡村迎来各项产业的建设,亦产生了不少乡村建筑,重塑了乡村精神文明。于是,许多建筑师被城乡精神文明的建立与重塑吸引,并渴望参与到这一过程中。在城市建设的改造层面,具体可聚焦工业厂房、历史建筑与街区、城市既有住区等项目的改造。建筑师需更多地面对城市的真实性,包括周边环境、邻里关系、城市肌理、新旧调和、街巷优化、运营管理等"即时具体的城市生活调整与文化再塑造"的问题。他们提出保留、拆除与延续的平衡策略,通过创造独特的空间体验,激发城市的新活力,并保持过去生活的纯粹性。在乡村建设的改造层面,具体可锁定在改造更新老宅或是对原有的村镇街区与地理景观进行整治更新。建筑师需更多地面对乡村的现实性,包括自然景观、村落活化、街巷优化、产业升级、运营管理、自发生长与外溢、身份认同、精神文明重塑、文化底蕴彰显等"深刻具体的乡村生活调整与文化再凝聚"的问题。而为了改善生活,乡村亦应首先重视保存、还原与流传,以及重新定义自我认同感,同时植入"外来"因素来重塑乡村风景,唤醒村庄。

参展建筑师中,张华、赵劲松、狄韶华、陈天泽、关英健、甄明扬、卜骁骏、张继元、张曙辉、王淼、申江海选定了"城市改造"板块;任军、郭海鞍、韩文强、田恬、王求安、张东光、刘文娟选定了"乡村改造"板块。他们从不同的视角解释了各自的改造领域,涉及的程度不一也揭示了所接触的人群性质及其背后运转机制的不同,更体现在策略、组群、关系、性质与价值上的差异与突破。

2019天津国际设计周"汇流而下的城乡之辩"建筑展是一项持续性的思辨哲学,同时让作品作为展示范本供外界注目、审视与讨论,并经由展览推出,希冀得到大家的反馈及鞭策。

As a key city in the Bohai Rim region, Tianjin is known for its rich and brilliant architectural culture. The heritage of historical buildings is a significant part of the architectural culture in Tianjin. Historical buildings such as Guangdong Guild Hall and the world-style buildings in the former Concessions in the old city merge with new buildings to shape the urban style that is not only a mixture of Chinese and Western elements but also an integration of classical history and modern trend, providing necessary assets and resources for the development of city, architecture and spatial environment of Tianjin. These historical buildings also play an important role in the process of urban renewal, and offer a crucial support for Tianjin's application for City of Design. In addition, under the architectural landscape of contemporary China, be it cities, villages, or between urban and rural areas, stock and increment are one of the prerequisites to face and consider before design. Therefore architects will start from this perspective and make use of their design wisdom to solve practical problems and satisfy practical needs of projects at different levels and angles.

As a result, in the Architecture Exhibition of 2019 Tianjin International Design Week, as the Chief Curator and Secretary of the Party Committee of the School of Architecture, Tianjin University, Prof. Song Kun conceived the theme together with his curatorial team members—Huang Yuanzhao, Zhang Xinnan, Hu Zinan, Zhao Wei and Li Dexin. The exhibition was themed on "Confluence", an embodiment of the architectural culture of Tianjin, to reflect the variables of the projects caused by stock and increment with the assembly of practicing architects from Beijing and Tianjin. The Architecture Session was organized in the forms of presentation of design works and face-to-face exchanges, attempting to create new thoughts. The curatorial team invited a total of 13 groups of architects.,who came from design institutes (Zhang Hua, Ren Jun, Guo Hai'an, Chen Tianze, Guan Yingjian, Zhen Mingyang and Tian Tian), designer firms (Di Shaohua, Wang Zhenfei, Bu Xiaojun, Zhang Jiyuan, Zhang Shuhui, Wang Miao, Wang Qiu'an, Zhang Dongguang, Liu Wenjuan and Shen Jianghai), universities (Zhao Jinsong and Han Wenqiang).

With "Confluence" as the theme of this exhibition and stock and increment as the background support, the curatorial team decided to focus more on the renovation of old buildings rather than new constructions, that is to say, to make moderate increment on the basis of the existing buildings, and pay attention to the aspects of renewal, activation and new addition. What is more, the curatorial team was determined to propose a dichotomous block theory in line with the development of the current architectural landscape, and divided the renovation into urban and rural sections. Consequently, two exhibition sections of "urban renovation" and "rural renovation" were arranged accordingly, which took shape respectively and formed two big branches to jointly converge in the Architecture Exhibition. The curatorial team plans to mold the Architecture Exhibition into two display sections, forming a temporarily opposite "virtual" state. Dialectical discussions were strengthened concerning the authenticity and reality of the dichotomy between the city and the country from the concept expression to the practice presentation. With the cross arguments the core value of the exhibition theme (Confluence) was clearer. After the two display sections are established, the participant architects selected their section of interest according to their design practice and the proportion of their participation in the renovation projects. The dialogue between the two sections thus revealed the debate between the city and the country, presenting the ultimate goal of the Architecture Exhibition in good faith.

The human settlements are divided into urban and rural regions. In China, urban and rural regions show their respective advantages through a series of construction projects. Based on the modernization construction after the Reform and Opening up, cities expand their land and carry out building projects with high efficiency, leading to the emergence of many urban buildings and establishment of the urban spiritual civilization. Meanwhile, thanks to the implementation of the rural revitalization strategy, rural areas ushers in the construction of various industries, generating a lot of rural buildings and reshaping the rural spiritual civilization. Therefore, many architects are drawn by the establishment and remodeling of urban and rural spiritual civilizations and eager to take part in the process. At the level of urban building renovation, the following project types may be targeted such as the renovation and renewal of industrial factories, old courtyard houses, historical buildings and blocks and traditional urban buildings. Architects need to be fully aware of the reality of urban area, including issues concerning "the instantaneous and specific adjustment of urban life and cultural reconstruction", such as surroundings, neighborhood relations, urban fabric, compromises between the old and the new, optimization of streets and lanes as well as operation and management. They propose a balanced strategy of reservation, demolition and continuation, which stimulates the new vitality of the city and preserves the purity of the old-fashioned lifestyle by offering a unique spatial experience. At the level of rural building renovation, the following projects may be focused such as the renovation and renewal of old houses or of traditional town blocks and geographical landscape. Architects need to be aware of the reality of rural area, including issues concerning "profound and concrete adjustment of rural life and cultural re-cohesion", such as natural landscape, village revival, optimization of streets, industrial upgrading, operation and management, spontaneous growth and spillover, identity, remodeling of spiritual civilization, and manifestation of cultural deposits. In order to improve the quality of life rural areas should firstly pay attention to reservation, restoration and circulation, and redefine sense of self-identity, while implanting "external" factors to rebuild rural scenery and activate the countryside.

Among the participatant architects, Zhang Hua, Zhao Jinsong, Di Shaohua, Chen Tianze, Guan Yingjian, Zhen Mingyang, Bu Xiaojun, Zhang Jiyuan, Zhang Shuhui, Wang Miao and Shen Jianghai chose the section of urban renovation; Ren Jun, Guo Hai'an, Han Wenqiang, Tian Tian, Wang Qiu'an, Zhang Dongguang and Liu Wenjuan selected the section of rural renovation. Both groups interpreted their respective sections from different perspectives. The different degrees of involvement also revealed the imparity in group of people and the operation mechanism behind it, which was more reflected in the discrepancy and breakthrough in strategies, groupings, relations, properties and values.

The Architecture Exhibition entitled "The Confluence: the Debate between Urbanity and Rurality", as the subdivision of 2019 Tianjin International Design Week is a sustainable speculative philosophy. Meanwhile, the design works on display offer models for appreciation, review and discussion in hope of receiving feedback and suggestions.

2019 参展建筑师及作品
Participant Architects & Their Works

玉庆成美术馆

张华 Zhang Hua

天津大学建筑设计规划研究总院有限公司/张华工作室
Tianjin University Research Institute of Architectural Design and Urban Planning Co., Ltd./Zhang Hua Studio

于庆成美术馆

于庆成是世界著名的原创泥塑家，风格独特，自成一家，蜚声中外。

于庆成美术馆的设计构思用感性的词形容就是"捏泥巴"，用理性的词形容就是"流形"。设计师对近年来的思考做了一个总结，提出了"流形"这一新概念，即一个从头至尾充满运动变化并有两个或多个不同表现形式端头的形体。该形体表现的不是一个结果而是一个过程，一个不断流动变化的空间形体，一个从静态到动态的过程，一个时空演变的过程，一个机体生长的过程，一个没有焦点的建筑，一个包含有从线性到非线性变化的几何构成，兼具拓扑与分形的特征。

于庆成美术馆的建筑形体空间具有十个变化：
力学——体态从静到动；
微分——形式从直到曲；
层级——分块从大到小；
光学——颜色从深到浅；
测量——面层从厚到薄；
计量——缝隙从宽到窄；
物理——质感从粗到细；
维数——空间从二维到三维；
性状——气质从刚到柔；
哲学——属性从阴到阳。

曾经习以为常的"方就是方，圆就是圆"的惯性思维被颠覆，这个不方不圆的形体不合常理但合乎逻辑。曲线与直线不再以对比的、机械的、刚性的欧式几何面孔出现，而是在从同胚、同伦到非同胚等一系列的拓扑变换下做非线性的变化。

Yu Qingcheng Art Museum

Yu Qingcheng is a world-famous original clay sculpture artist with a unique style, who is renowned in China and oversea.

To describe this design, an emotional phrase is "playing with mud", a rational word is "manifold". The designer made a summary on thinking in recent years and put forward a new concept of "manifold", which is a shape full of movement and with two or more different forms at different ends from the beginning to the end. It is not a result but a process, a continuously flowing and changing physical space, a process from static to dynamic, a temporal and spatial evolution process, a body growing process, a non-focus building, a geometrical composition from linear to nonlinear changes with both topology and fractal characteristics.

Physical space of Yu Qingcheng Art Museum has ten changes:
Mechanics—the shape from static to dynamic;
Differential—line from straight to curve;
Level—block from large to small;
Optics—color from dark to light;
Measurement—surface layer from thick to thin;
Meterage—gap from broad to narrow;
Physics—texture from coarse to smooth;
Dimension—space from two-dimensional to three-dimensional;
Character—temperament from rough to soft;
Philosophy—attributes from Yin to Yang.

The conventional mode of thinking that "A square is a square and a circle is a circle" is overturned. Neither square nor circular, the present object is unconventional but logical. Curves and lines are no longer presented in the form of contrastive, mechanical and rigid Euclidean geometry shapes, but make nonlinear transformations under a series of topology changes from homeomorphism and homotopy to non- homeomorphism.

于庆成美术馆

战争纪念馆

战争纪念馆

在宁静的长河中,壮观的瀑布和湍流在地质突变中出现;

在山脉中,怪诞的山峰出现在地质断裂的碰撞中;

在漫长的一生中,不可预知的人生转折将改变一个人的命运。

在人类历史上:

世界之战——人类的自我扭曲;
家国之争——人类的自我纠缠。

War Memorial

In the serene long river, spectacular waterfalls and turbulence appear in the geological mutation;

In the mountain range, grotesque peaks occur in the collision of geological rupture;

In a long life, unpredictable turns of life will change one's fate.

In the history of mankind:

The world wars are the self-distortion of mankind;
The family-country fights are the self-entanglement of mankind.

赵劲松
Zhao Jinsong

天津大学建筑学院/非标准建筑工作室
School of Architecture, Tianjin University/Non-standard Architecture Studio

山东农业大学改扩建工程

此处乐，不思乡——一个奇葩校园的奇葩改造。

通过一系列围绕农业的"奇葩"操作，将原有校园改造成史上最"农"的大学校园，从而重新塑造人们对"农"的固有看法，激发师生对"农"的自信认同，唤醒"农"所蕴藏的时尚潜力，让新的校园与"农"一起迎接更好的未来。

乡愁不必在远方，此心安处才是家。

The reconstruction and expansion plan of Shandong Agricultural University

Homesickness Healer—A weird campus reconstruction plan beyond your imagination.

Through a series of creative architectural design methods for agriculture, the original campus is reformed to the one that can best reflect the characteristics of agricultural universities. On the one hand, this plan will redefine people's views on agriculture and inspire teachers and students to be more confident on agriculture and accept it better. On the other hand, it will awaken the hidden fashion potential in agriculture and make the new campus together with agriculture embrace a better future.

Homesickness is not necessarily far away, no matter where you are, as long as it is a place of inner peace, it is hometown.

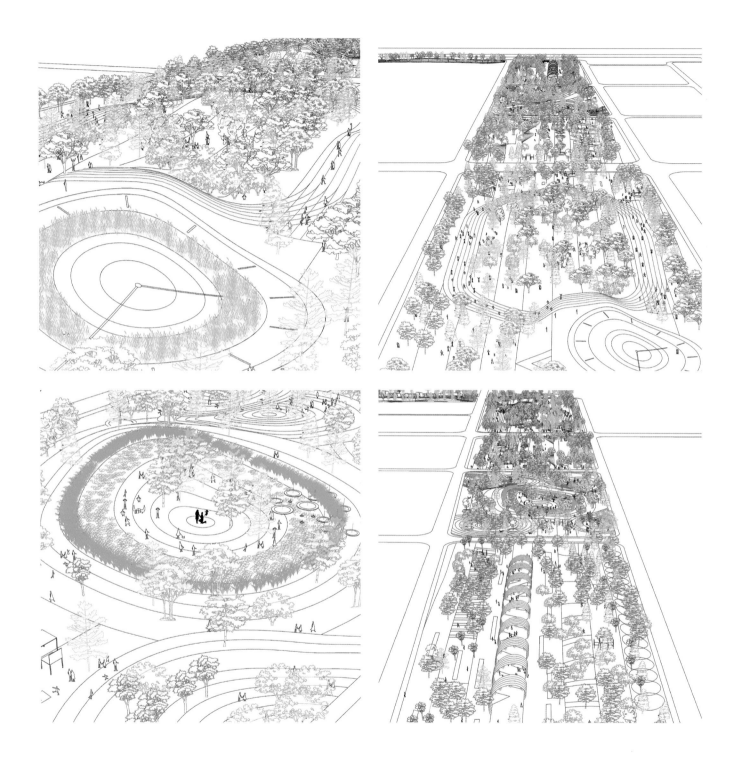

天筠工业办公组团

狄韶华
Di Shaohua

第一实践建筑设计
PRAXiS d' ARCHITECTURE

第一实践建筑设计事务所承接了一系列尺度不一的项目:从办公建筑、艺术家工作室、展览空间到装置、家具等设计。第一实践的设计始于对基地的认知,从其物理环境和文化环境中寻找灵感,然后获取一个主题。主题将通过建筑的媒介得以呈现,主题决定材料、光线和空间的特质。在与基地条件、项目实际要求以及其他外界的限制因素相互碰撞的过程中,主题被具象地体现出来。建筑设计就是在这个"碰撞"过程中,去发现主题被感知和不被感知的界线。这个过程是一个对比例的权衡,在真实与想象之间,实际与主题之间。有的时候主题能够被转化成建筑的空间经验,有时则不能,那么也许应该重新获取主题,因为建筑具有实用性。

此外,第一实践建筑设计事务所执着于通过对项目基址人文和物理环境特质的认知,使建筑与环境巧妙地融合在一起,并利用经济、环保、本地易获得的材料和建造技术,创造出不寻常的建筑空间体验。不论项目尺度大小,均最大程度地赋予建筑以实用性和精神性,并力求精益建造。

PRAXiS d' ARCHITECTURE is actively engaged in a broad range of projects of various scales, from offices, artist studios, exhibition spaces to installation, furniture, etc. The design of PRAXiS d' ARCHITECTURE begins with identification and analysis of cultural and physical context of the site and drawing inspiration from it. Then a theme of design is to be acquired. This theme will be tangibly represented through the device of architecture, and determines the nature of material, light and space. The theme will be transformed as a result of the reaction when it confronts with site, programmatic requirements, and other external forces. Design process is to discover, during the "reaction", the threshold between perceptibility and imperceptibility of the theme. This process is calibration of proportion between the real and the mental, the fact and the theme. Sometimes the theme can be resolved into architectural experience, but sometimes it will never be resolved unless the theme is reacquired, because architecture bears practicality.

In addition, PRAXiS d' ARCHITECTURE adopts an ingenious and harmonious approach to integrate architecture with its surroundings based on the understandings of the site's cultural and physical context. It strives to create extraordinary spatial experiences by using economized, eco-friendly, local materials and available means of construction. Regardless of project scale, it seeks to fully realize architecture's practical values by lean construction.

怀斯《克里斯蒂娜的世界》之隐居

任军
Ren Jun

天津市天友建筑设计股份有限公司
Tianjin Tenio Architecture and Engineering Co., Ltd.

绿色乡村
乡村农宅:从低能耗到零能耗

通过大兴半壁店村的两个乡居案例——"陋居"和"零舍",建筑师探索了乡村居住建筑的绿色策略。

"陋居"以零煤耗为出发点,从能源形式、室内舒适性、乡土材料等方面尝试了低能耗乡居。"零舍"结合国家《近零能耗建筑技术标准》,以性能化设计方法将被动式超低能耗与太阳能光伏瓦、光伏玻璃相结合,用装配式模块创造出乡村零能耗居住产品。

乡村产业:从"三产文创"到"多产院落"

在大兴半壁店村引入泰迪熊知识产权,将文创产业引入乡村。泰迪熊博物馆采用被动式太阳房,在屋顶上种满了当地的月季花,并以玉米遮阳幕墙彰显乡村特色。而在昌黎葡萄沟的山村,用可循环的建筑空间将第一产业种植、第二产业加工、第三产业消费融合在一起,实现了微影响与零排放。

Green Countryside
Rural Farmhouse: From Low Energy Consumption to Zero Energy Consumption

This design explores the green strategy of rural residential buildings through two cases of farmhouse in Banbidian Village, Daxing District, which are "The Burrow" and "Zero Cottage".

Starting from zero coal consumption, "The Burrow" tries low-energy rural residence from the aspects of energy form, indoor comfort and local materials. According to *Technical Standard for Nearly Zero Energy Buliding* in China, "Zero Cottage" combines passive ultra-low energy consumption with solar photovoltaic tiles and glass by performance-based design method, and creates a rural zero-energy residential product with prefabricated modules.

Rural Industry: From "Third-industry Culture and Creativity" to "Multi-industry Courtyard"

Teddy Bear IP was introduced into Banbidian Village, Daxing District so as to introduce the cultural and creative industry into the countryside. The Teddy Bear Museum, based on a passive solar house, has a roof covered with local Chinese roses and a corn sunshade wall to show the characteristics of the countryside. In Putaogou, Changli County, the recycling architectural space integrates the primary industry (plant), secondary industry (processing) and tertiary industry (consumption) to achieve micro impact and zero carbon emission.

米勒《拾穗者》之零舍

梵高《收获的麦田》之泰迪熊博物馆　　托马斯·科尔《帕特南堡的景色》之昌黎多产院落

137

前洋村农夫集市庭院

郭海鞍
Guo Hai'an

中国建筑设计研究院乡土创作研究中心
Rural Culture D-R-C, China Architecture Design & Research Group

中国新乡土解译

当代中国，无论城市或乡村，都面临着风貌趋同、千城/村一面、特色丧失的窘境。

通过尝试以乡土文本和解释为线索的转译方式，将乡土属性和情境进行输出，从而再现中国传统乡土文化的本质，这对于衔补中国建筑的脉络非常重要。即便文本信息可能局部缺失甚至扭曲，但是其基本的定义和属性还能在现象中得到基于新乡土立场思考的解释和转译，保持和假定一些本土特质，使得中国城市与乡村存续自身的特色，使得当代建筑设计的发展无愧于历史与未来。

Interpretation of New Vernacular in China

In contemporary China, whether cities or the villages, are all faced with the dilemma of the convergence of styles, the assimilation of appearance and the loss of characteristics.

By trying a translation method with local texts and explanations as clues to output local attributes and situations, the design obtains an essential reproduction of traditional Chinese vernacular culture, which is significant to integrate and complement the vein of Chinese architecture. Even if the text information may be have partial loss or even distortion, but the basic definitions and properties still can obtain interpretation and translation based on the new vernacular position in the phenomenon, and maintain or assume some local traits so that Chinese cities and villages can keep their own characteristics, the development of contemporary architectural design is fit for the history and the future.

前洋村农村集市

九峰村乡村客厅

西浣村昆曲书社

九峰村乡村客厅室内

扭院儿

韩文强
Han Wenqiang

中央美术学院建筑学院/建筑营设计工作室
School of Architecture, Central Academy of Fine Arts/ARCH STUDIO

本次旧城更新以散落于胡同的单个院落为切入点，对其进行针灸式的改造，试图结合新型业态和场景体验激发旧城活力，让空间成为不同社群新与旧、内与外之间的媒介。

曲廊院
曲廊院是在旧有建筑的屋檐下加入一个扁平"曲廊"，将分散的建筑合为一体，创造一种新旧交替、内外穿越的环境感受。

扭院儿
扭院儿尝试改变四合院给人的庄重、刻板的印象，营造一种开放和活跃的生活氛围。根据现有的庭院布局，利用起伏的地面连接室内外空间，并延伸至房屋内部扭曲成为墙壁和屋顶，让内外空间产生新的动态关联。

叠院儿
叠院儿受到"叠合院落"的启发，将原本的内合院改变为"三进院"，以此逐步适应从公共空间到私密空间的递进，并利用院落的逐层过渡在喧闹的胡同街区之中营造出宁静、自然的诗意场景。

折叠院
折叠院设计最大限度地保留已有建筑的肌理，同时利用连续起伏的楼梯将平屋顶与地面相互串联，形成一条可达可玩的立体折叠的景观环线，让建筑的内与外、上与下、新与旧产生新的互动。

The old city renewal takes individual courtyards scattered in hutongs as the breakthrough point to carry out the transformation of acupuncture style. The design intention is to inspire the vitality of the old city in combination with the new format and scene experience and make the space a medium between the new and the old, the internal and the external of different communities.

Curvy Corridor Courtyard
Curvy Corridor Courtyard is a flat "curving gallery" added under the eaves of an old building, integrating the scattered buildings into one and creating a sense of environment where the old and the new alternate and pass through inside and outside.

Twisting Courtyard
The design of Twisting Courtyard aims at getting rid of the solemn and stereotyped impression given by Siheyuan (quadrangle courtyard) and creating an open and active living atmosphere. Based on the existing layout of the courtyard, the undulated floor is used to connect indoor and outdoor spaces. And it is extended to the inside of the house, twisting into walls and roof, thus creating a dynamic connection between inside and outside.

Layering Courtyard
The design of Layering Courtyard inspired by Multiple Layering Courtyard changed the previous inner courtyard into a Three-layered Courtyard in order to adapt to the transition from the public space to the private space step by step. Besides, the designers took advantage of the yard's layering structure to create a quiet, natural and poetic scene in a noisy Hutong neighborhood.

Overlapping Courtyard
Overlapping Courtyard design retains the texture of the existing building to the utmost extent, and uses the continuous and undulating stairs to connect the flat roof and the ground to form a three-dimensional folded landscape loop that allows access to, creating new interactions between the interior and the exterior, the upside and the downside, the new and the old.

叠院儿　　折叠院

陈天泽
关英健
甄明扬
Chen Tianze
Guan Yingjian
Zhen Mingyang

天津市建筑设计研究院有限公司
Tianjin Architecture Design Institute Co., Ltd.

天拖故事
"天拖"是什么？

但凡在天津长大的人都可能说上几句："天拖就是天津拖拉机制造厂。""天拖是天津的骄傲，毛主席、周总理都来视察过。""天拖是相声《纠纷》中'丁文元'的工作单位……"

随着城市的发展变迁，天拖工厂早已搬离市区。大面积的厂房是拆掉还是保留改造？它们应该以什么样的姿态融入城市？我们还有没有机会延续历史与城市发展的记忆？再过10年，孩子们是否还能说得出天拖的故事？

2012—2013年，由规划部门牵头，天津市建筑设计研究院有限公司、天津市筑土建筑设计有限公司协同工作，对天拖地区进行城市设计及工业遗产更新专项规划。规划将天拖地区建设成集时尚消费、科贸创意、生态宜居于一体，可以体现天津工业历史风貌的区域中心。

对于有识别特征的空间和构筑物，方案采用加固、翻新的方式保留其原有特点，再通过适当的手法体现功能置换后新的空间特征，在保留与改建、新建，历史真实性与建筑适用性中取得平衡。由此，在新天拖区域中，新建建筑与保留的历史建筑共生共存，相互联系又彼此区分。天拖的历史印记被保留下来，天拖故事也被印刻到城市发展的肌理深处。

The story of Tian Tuo
What is "Tian Tuo"?

Ask anyone who grew up in Tianjin, you may get such answers as follows: "Tian Tuo is Tianjin Tractor Factory." "People in Tianjin are all proud of Tian Tuo, which was inspected by Chairman Mao and Premier Zhou." "Tian Tuo is Ding Wenyuan's work unit in crosstalk *Dispute*."

With the development and change of the city, Tian Tuo has to move outside the center of Tianjin. Should the left factory buildings be demolished or reformed? How should it integrate into the city? Do we have a chance to tell its everlasting story to the future generations? In another ten years, what answers can we get from the younger generations?

From 2012 to 2013, led by the planning department, Tianjin Architecture Design Institute Co., Ltd. and Archiland (Tianjin) Architectural Design Co., Ltd. worked together to make a special plan of urban design and industrial heritage renewal for Tian Tuo area. Tian Tuo area was planned to transform into a regional center integrating fashion consumption, science and trade creativity with pleasant living environment, as well as reflecting Tianjin's industrial history.

For the space and structures with identification features, the design used appropriate methods to characterize the space after its function had been changed. The balance between preservation and reconstruction as well as new construction, historical authenticity and architectural applicability has been achieved. Therefore, in the new Tian Tuo area, new buildings and historical buildings coexist, interact and differentiate with each other. The historical spirit of Tian Tuo has been preserved, and the story of Tian Tuo has also been engraved into the development of the city.

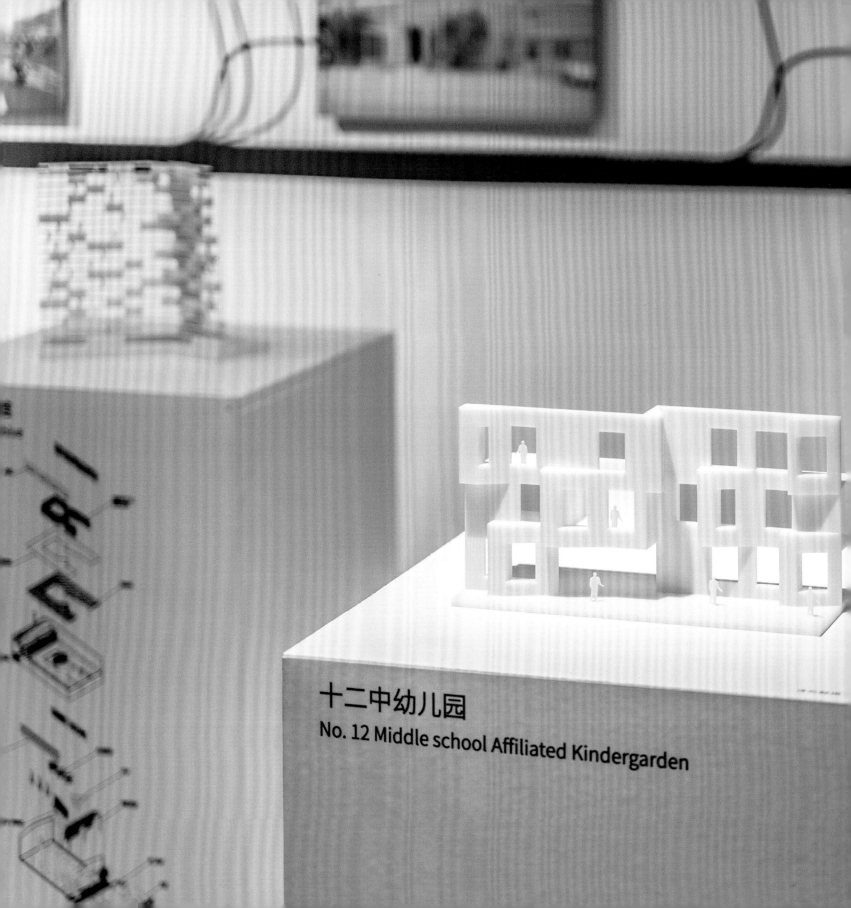

十二中幼儿园
No. 12 Middle school Affiliated Kindergarden

卜骁骏
张继元
Bu Xiaojun
Zhang Jiyuan

时境建筑设计事务所
Atelier Alter Architects PLLC

再创历史——时境建筑的历史观图景

近年来建筑市场合理化后，中国建筑不再像10年前那样依赖于英雄主义的新建筑；激烈的市场竞争促使人们对建筑的思考更加理性，引发了对现有建筑更新改造设计的巨大需求。身处在这个洪流当中，作为建筑实践者的时境建筑没有脱离整个大环境，而是积极参与并思考这种特殊的物质条件。

相比于欧洲等地区，中国有着不同的建筑条件，比如过去一些建筑的质量并不理想或者不具有历史文脉特征；或者一些建筑具备文脉特征但质量不理想，由于土地政策等原因，拆掉后难以重建；有些由于现有的结构依然可以使用，仅做外部更新。但这中间往往掺杂着对功能的重新构建，时境建筑在面对这些问题时，也需要考虑如何处理多方面的需求。

时境建筑在中国的9年实践中，建筑改造类的项目已从开始的20%上升到近年来的50%甚至60%。面对这个重大的建筑课题，时境建筑该采取怎样的态度对待历史、对待既有的物理环境呢？对于情况非常好的、结构保存完好的改造对象来说，是像之前一样修建旧的部分还是自命不凡地对其"暴改"呢？

时境建筑坚持以分析的态度对待每一个建筑的既有条件，既不墨守成规，又不完全割裂历史。对待历史，建筑师应该有自己的创新思想，在产生新建筑语言的同时还要尊重历史。

经过对以往工作的整理，时境建筑形成了五个层面上不同力度的设计点，分别是功能、结构、流线、材料和细部。通过整理，自然地形成了一张针对历史进行物理操作的宏观图景，在这个图景下时境建筑开启了对重构历史的思考，展示了时境设计历史观的力量。

Reconstructing History—Atelier Alter's Historical View works

After the rationalization of the construction market in recent years, Chinese architecture no longer shows its dependence on the heroic new architecture as it did 10 years ago. Fierce market competition promotes more rational architectural thinking and brings a huge demand for the renovation design of existing buildings. In the midst of this boom, as a practitioner of architecture, Atelier Alter doesn't disconnect from the overall environment, but actively participates and thinks about this special physical condition.

Compared with Europe and other regions, China has different construction conditions. For example, some buildings in the past was not ideal in the quality or featured by historical context. Or, some buildings featured by historical context but not ideal in the quality, because of the problem of land policy, are difficult to rebuild after demolition. Since the existing structures of some buildings are still usable, only external renewal is needed. However, this is often mixed with the reconstruction of function. When faced with these problems, Atelier Alter also needs to consider how to deal with various needs.

In the 9 years of practice in China, Atelier Alter's renovation projects have increased from 20% at the beginning to 50% or even 60% in recent years. When facing this significant architectural issue, what attitude should Atelier Alter take towards history and existing physical environment? For those in good structural quality, do we repair the old as before or change everything?

Atelier Alter holds the attitude of studying the existing conditions of each building, which neither sticks to conventions nor completely disconnects from the history. Towards history, architects should have their own creative ideas in dealing with history and show respect for history while generating new architectural language.

After archiving the past works, Atelier Alter have formed the language at five

levels, with different emphasis, which are programs, structures, circulations, materials and details. Through these arrangements, a macroscopic picture of physical operation of history is naturally formed. Under this picture, Atelier Alter start to think about the reconstruction of history and demonstrate the power of historical view of Atelier Alter.

田恬
Tian Tian

天津市城市规划设计研究总院有限公司
Tianjin Urban Planning & Design Institute Co., Ltd.

做"看不见的设计",唤起对乡村景观的认同——天津西井峪村景观营造实践

西井峪又名"石头村",是位于天津市蓟州区的一座历史文化名村,具有传承乡村文化景观的特殊使命。同时,它同大多数乡村一样面临经济萎靡、发展迟缓、生态恶化、风貌衰落等问题。作为设计师,如何避免以惯用的技法和过于主观的臆想来判断乡村的发展进程?如何避免一味地追求自我表达而过度设计?如何避免"破坏性建设"的种种尴尬?什么才是"景观设计下乡"的真正意义?这些问题贯穿了西井峪景观营造的4年实践过程。

西井峪地形复杂,土壤贫瘠,水源也不稳定,是一个不易生存的地方。村民用一些生产、生活的方式来应对和破解环境难题,呈现出地尽其利、物尽其用、人尽其能的地域性景观。最突出的体现就是浅层地表石材的在地利用,形成依形就势、色彩斑驳的石头村。乡村的风景没有刻意设计的形态,一切都与生活保持着最直接和最紧密的联系。从民居单元、村庄聚落以及承载聚落的自然地理环境中发现人地关系的规律,挖掘西井峪村的景观特征,是此次景观营造工作的基础。

在西井峪村,最具特色的风景就是它的宅基地单元石砌边界景观。宅基地边界既是村民居住空间的分界,也是村落中最为生动活泼的风景线。各家院落有大有小、形状各异,其封闭程度也各不相同,这些组合随机应变、因地制宜且不拘一格。虽然建筑单体形式大同小异,但居住单元一经与特定的地理、地貌、区位相结合,便产生了千姿百态、非人工所能达到的丰富变化。这也反映了传统乡村的共性:地域自然资源的特质常常决定了结构方法,结构方法又大大约束了表现形式,因而一个贯穿始终的建造逻辑使得村庄风貌和谐统一;针对每一个局部的处理又有一个局部的解决办法,最终形成具有丰富变化的集合。

制定规则,用"石头与食物"做设计,即以本地石材和本地作物为建造材料。其原因如下:将设计融于无形,让人身在其中却感受不到设计的痕迹;使用本地材料,严格控制造价,并且将景观维护的难度和成本降到最低;使用当地熟悉的建造工艺,能够使更多村民参与到建设中来,有助于本地传统工艺的延续和发展;使新的建筑融入村庄传统建造规律之中。

合作设计,形成聚力。邀请村民参与,倾听村民心声,让设计更接地气儿。以"地尽其利、物尽其用、人尽其能"为建设原则,最大化利用乡村资源,包括物质资源,也包括村民的智慧和经验。例如,与村民一起合作修缮农户宅基地边界,改造村中心广场,修缮排水系统,建造山地停车场,修建公共休憩设施,修整道路与街巷等。同时,建筑师团队对民宿、书店、游客服务中心等示范性建筑的改造,以及视觉设计师团队为西井峪乡村视觉设计所做的工作也很好地示范了如何立足传统,用新观念、新技术为西井峪村带来高品质发展。乡村运营团队积极培育有西井峪特色的产业经济。生活品质提高,游客络绎不绝,让西井峪村民看到了保护好传统物质空间带来的实际利益和发展机会。

做"看不见的设计",唤起村民对乡村景观的认同。"高高山顶立,深深海底行",乡村工作既需要设计师在职业上的专业性和远见,也需要在乡村营造的点点滴滴中躬耕前行。在西井峪乡村实践的4年,设计师最重要的价值就是改变了村民对于景观的理解,让村民认识到"砌石头"带来的美学价值、文化价值和发展机会。"把西井峪建设得更像西井峪"成为设计师与村民达成的共识。

Make "Invisible Design" to Arouse the Recognition of Rural Landscape — Landscape Practice of Xijingyu Village in Tianjin

Xijingyu, also called "Stone Village", is a historical and cultural village situated in Jizhou District, Tianjin, with a special mission of inheriting cultural landscape in the village. At the same time, like most villages, it faces such problems as flagging economy, lagged development, deteriorated ecological conditions and declined styles and features. How can designers avoid judging the development progress of the village with idiomatic techniques and excessively subjective assumption? How to avoid excessive design due to excessive pursuit of self-expression? How to avoid embarrassment caused by "destructive construction"? What is the real significance of "landscape design in villages"? These questions go through the four-year landscaping practice in Xijingyu.

With complicated terrain, poor soil and instable water source, Xijingyu is not an environment suitable for survival. Villagers deal with environmental problems by means of some ways of production and life, showing a regional landscape of making the best use of everything. The utilization of shallow surface stones is the most prominent embodiment, which forms a stone village with various colors according to the landform. Many landscape elements in villages do not have deliberate forms, and everything has the most direct and the closest relation with life. Finding rules in the relation between people and geology from residence units, village settlements and natural geological environments of the settlements and digging landscape features of Xijingyu Village are the foundation of landscaping this time.

The stone boundary landscape of house site units is the most featured landscape in Xijingyu Village. The boundary of the house sites is both the boundary of the living space of villagers and the most lively and vivid landscape in the village. Various courtyards have varied sizes and shapes as well as different close extent. Such combinations adapt themselves to the changing circumstances and follow no set form. Although the building units have almost the same form, there are abundant changing forms once residence units are combined with specific geological conditions, landform and locations. It reflects the generality of traditional villages. The characteristics of regional natural resources usually determine the structure method, which restrains the expression form greatly. Therefore, an unchanged construction logic leads to a harmonious and unified landscape of the village. In addition, local solutions are provided for local treatments, forming a collection of abundant changes.

Making Regulations and Designing with "Stones and Foods". That is to say, local stones and crops are used in the construction for the following reasons: firstly, design dissolves in an invisible way to make people feel no trace of design; secondly, the use of local materials can control the construction price strictly and reduce the degree of difficulty and the cost of landscape maintenance to the minimum; thirdly, adopting local mature construction techniques can make more villagers participate in the construction, which contributes to the continuity and development of local traditional techniques; fourthly, new buildings can be integrated into traditional building rules of the village.

Designing Cooperatively for Resultant Force. It is needed to listen to the voice of villagers and to invite them to participate in the design so as to make it "closer to villagers". It is needed to make full use of resources in the village according to the construction principle of "making full use of everything", including both material resources and the intelligence and experience of villagers. For example, cooperating with villagers to repair the boundary of the house sites, transforming the village center square, repairing the water drainage system, constructing the mountain parking lot, building the infrastructure and improving the roads and streets. At the same time, the architect team's transformation of demonstrative buildings such as dwellings, book stores and tourist service centers and the visual designer team's village visual design show the way to facilitate high-quality development of Xijingyu Village with new concepts and technologies. The village operation team cultivates the industry economic vigor with features of Xijingyu. With the improved living quality and the increasing number of tourists, Xijingyu Village shows the actual benefits and development opportunities brought by preserving traditional materials and space to villagers.

Make "Invisible Design" to Arouse the Recognition of Rural Landscape ."We should have a great pattern of aspiring and cultivate a down-to-earth attitude." Designers are required to have professional skills as well as foresight and take effort to build villages. During the four-year practice in Xijingyu Village, the most great contribution designers have made is changing the villagers' ideas about landscape and making villagers realize the aesthetic value, cultural value and development opportunities brought by "stone masonry". "Constructing Xijingyu as a village with more local characteristics" becomes the consensus of the designers and the villagers.

张曙辉
王淼
Zhang Shuhui
Wang Miao

北京八作建筑设计事务所有限公司
Beijing Bazuo Architecture Office, Ltd.

加减
"+"1lx：拉萨市人民中心医院
自然光线的重要性
根据不同人群的活动和使用需求，在相应区域都引入了自然光线。

1.普通地下车库一般在地下二层，无自然采光。根据西藏地域广阔的特点，将地下室设置在负一层，能更方便地引入光线。

2.全院采用防雨采光窗，使整个地下室实现自然采光，完全满足使用要求。整个地下室全年无采光能耗，每年节省500万元的开支。

"－"1m²：建平县红十字会医院
通过分析人群的使用习惯及行为活动，减少不必要的1m²，去优化必要的1m²，使空间品质上升。

1.1个医生一天最多能看40个病人，1个诊区最少需要多少个医生？因为医生错峰上班，我们建议2个医生共用一间诊室，一个诊区总体诊室面积可缩减1/3。

2.门诊医护区、诊室、医技医护区、病房区、行政楼等都是医护人员区域，空间重合、浪费（门诊医生90%时间待在诊室等）。

3.缩减独立办公，改成集中办公（减少面积）。

Addition and Subtraction
"+" 1lx, Lasa Central Hospital
The Importance of Natural Light
Based on the activities and needs of different people, Bazuo introduces natural light to corresponding regions in the design.

1. Natural lighting always cannot reach the underground garage on the second floor underground. SinceTibet,has a vast area, the basement is set on the first floor underground so that nature light can enter the room more easily.

2. Natural lighting is enough for basement by the rainproof windows on the ground. This could save 5 million RMB for each year.

"－" 1m², Red Cross Hospital in Jianping County
In the design, Bazuo reduces every unnecessary 1 square meter, and optimizes the necessary 1 square meter through the analysis of the habits and behavior of people.

1. One doctor receives 40 patients a day, what's the minimum number of doctors needed in one area? Considerating staggered commuting, we suggest two doctors share one consulting room so that consulting rooms are reduced by 1/3.

2. There is space overlapping and waste (doctors spend 90% of time in consulting room) among staff area in Outpatient area, consulting room, D&T area, ward area, administrative building and so on.

3. Independent offices are reduced and changed to centralized offices (to reduce area).

湖南邵阳清水村乡村振兴项目

王求安
Wang Qiu'an

北京安哲建筑设计有限公司
ANT Architectural Design Co., Ltd.

河北阜平大柳树村规划设计

大柳树村的整体规划充分尊重了原有的台地肌理，做了树状的规划设计，并将村庄北侧的原拆迁村设计为遗址公园，保留了老村的变化痕迹。

设计中采用了复式户型设计，利用了场地既有的6m高差的台地。一、二层的居民可以由南侧直接进入，三、四层的居民则由北侧的台地进入。由此，村民可以依旧延续在平房的居住习惯，不会产生被迫上楼的抵触心理。在每一户的院子中，方案设置了大量的储藏空间用以存放农具，在尊重村民原有生活习惯的同时，为他们提供更多的便利条件。

村中设置了一所学校，含幼儿园和小学。为了帮助小朋友释放自己的天性，方案在设计时有意地增强了学校的趣味性。在学校的中庭处，以莫比乌斯环为概念设置了一条一直延伸至屋顶的坡道。小朋友们可以在这里自由奔跑，探索这个有趣的世界。希望这个有趣的学校在孩子们的早期教育中起到积极作用，使他们成为追求美好、对世界充满求知欲的人。

湖南邵阳清水村乡村振兴项目

村庄的整体设计本着尽量避免大规模拆建这一原则，沿场地一侧布置了村部，以便退让出尽可能大的广场作为村民的活动场地。清水村气候湿润，雨天较多，建筑设置了层层叠叠的屋檐，这在雨天将呈现出一种别样的景致。

在清水村，由青砖、土坯砖砌筑的老房子不多见，但正是这些老房子体现了乡村的韵味。于是，在村部的设计中，方案以村中拆除的旧青砖、土坯砖为主要材料，希望通过使用这种朴素的建材，在向村民展示材料魅力的同时，让那些居住在老房子中的人们重拾信心。村部的整体设计与村中其他建筑相适应，仿佛是从这片土地中生长出来的一般。在村部的空间设计上则注重游览系统的规划，大量的连廊、楼梯、露台等被设置，使村民在雨天亦有属于自己的活动场地，这在多雨的湖南地区十分重要。此外，村部还可作为村民图书馆，成为村民及小朋友们交流的重要场所。

湖南西湖管理区工人文化宫

工人文化宫所处的基地是湖南地区少有的平原，是西洞庭湖边的一片沃土。这里的居民多是40多年前从新化和安化移民而来的，他们既是农民也是农场职工。他们通过自己的艰辛耕耘，将这里改造成如今美丽的家园，这背后除了艰辛的劳动外，更有着无法消散的乡愁。

在这里，方案为他们打造了一个美好的圆形岛屿，满植桃树，浪漫而亲切。一个好的公共建筑可以改善人们的生活，随着项目的落地，这里人们的娱乐活动由原本单一的打牌喝酒变得更加多元化，更多的人爱上了喝茶与读书。

在借鉴了梅山地区村落建筑元素的基础上，方案试图重新组织这个多功能的职工活动中心，使其成为活动聚落，并采"当代"的手法设计院落、街道、巷子、码头等场景，借此空间氛围纾解村民们思乡的愁绪。

湖南常德西湖管理区洞庭渔村规划设计

渔村是"村"的一种特别存在模式，在村落群体中高度个性化。自然存在的形态是渔村文化的体现，因此设计在充分了解当地文化的前提下，深入挖掘这种形态特征，创造了一个村落群体式的新渔村。

作为西湖管理区的全区域规划设计及建设国家农业公园的重点项目，设计从布局、文脉、推进模式等方面进行了多种尝试，希望"新"村融入自然，在充分体现梅山文化和渔村特色的同时，为村民提供真正高质量的生活方式，让村民享受到大自然的美好。

Planning Design for Daliushu Village, Fuping, Hebei

With full respect for the original platform texture, a tree-shaped overall planning is made for Daliushu Village. The original village in the north is re-designed as a relic park where the changes of the old village are retained.

The duplex house design is adopted on basis of the existing platform with 6m height difference. The residents on the first and second floors can directly enter their houses in the south, and those on the third and fourth floors may enter from the north platform. Thus villagers can still keep their habits of living in one-storey houses, without resistance to go upstairs. In the courtyard of each house is plenty of storage space for store farm tools. This design has shown respect to the original living habits of villagers and provides them with greater convenience.

A school including kindergarten and primary school is built in the village. To help children liberate their nature, enjoyments are enhanced in the design of the school. In the atrium of the school is a slope extending to the rooftop, which is inspired by the Mobius band. Children can run freely and explore this funny world. This funny school can play an active role in children's early education and help them grow into persons who pursue perfection and are curious about the world.

Rural Revitalization Project of Qingshui Village, Shaoyang, Hunan

Guided by the principle of no massive demolition and construction, the Village Hall is built along the site to give as much as space for villagers' activities. In light of the moist climate and the great number of rainy days, the overlapping eaves are designed for the buildings to present a unique scene on rainy days.

The old houses built with black bricks and rammed earth blocks are not commonly seen in Qingshui Village, yet they exude the charm of the village. Therefore, the old black bricks and rammed earth blocks demolished in the village are used as main materials to re-build the Village Hall. Through the use of these plain materials, the design hope to restore their charm and help the dwellers of these old houses regain confidence. The Village Hall is perfectly blended into the other buildings in the village, as if grown from this land. The space design of the Village Hall highlights the planning of sightseeing system. Extensive corridors, staircases and terraces are designed to provide villagers with activity venues on rainy days, which is highly important in the rainy area of Hunan. Furthermore, the Village Hall also serves as the Village Library, a vital exchange venue for both villagers and children.

Workers' Cultural Palace, Xihu Administrative District, Hunan

The base where the Workers' Cultural Palace is located is a plain rarely seen in Hunan. It is fertile land along the West Dongting Lake. Local residents are mostly migrants who moved from Xinhua and Anhui about four decades ago. They are both farmers and farm workers. Through their hard work, they have turned this place into a beautiful homeland today. However, besides the hard work, behind it is also their nostalgia for their hometown that permeates in the air.

Therefore, the design builds a romantic round island fully planted with peach trees for them, romantic and intimate. A good public building can improve people's life, with the completion of the project, the design diversify their recreational activities and make more people love tea tasting and reading rather than card playing and wine drinking.

On the basis of borrowing the architectural elements of villages in Meishan area, the design attempts to re-organize this multifunctional staff activity center and make it an activity cluster. "Contemporary" methods are used to design scenes such as courtyards, streets, lanes and wharves to relieve the melancholy of homesickness.

Planning Design for Dongting Fishing Village, Xihu Administrative District, Changde, Hunan

Fishing village is a special mode of village, highly personalized in the village community. A form that naturally exists reflects the culture of fishing villages. Therefore, we have dug deep to explore the features of this form in this design after a full understanding of local culture and created a community-type new fishing village.

As a part of Regional Planning Design of Xihu Administrative District and a key project of National Agricultural Park Construction, this design has made multiple attempts in layout, context and progression pattern. We hope to blend the "new" village into nature in full expression of Meishan culture and features of fishing villages and provide villagers with a truly quality lifestyle and an opportunity to embrace the beauty of nature.

驿道廊桥改造

张东光
刘文娟
Zhang Dongguang
Liu Wenjuan

合木建筑工作室
Atelier Heimat

置身于一个传媒视觉化的时代,建筑师的实践成果通常以图像的方式被传播、消费,而生成建筑作品的动机、策略、方法及其背后的社会意义,相较而言容易被忽视。某种程度上,建筑师应该保持批判性的思维。

建筑师在设计工作中,除了需要对社会、环境、城市、乡村进行观察外,还需要综合掌握远超出专业范畴的知识,从自己的视角发现或者提出问题,以实践的方式进行回应与探讨。唯有这样才能立足当下,产生具有根本性创新的设计。

合木建筑在本次展览所呈现的四个微型项目(合阳半宅、驿道廊桥改造、骆驼湾桥改造、声之穴),从建筑学的基本问题出发,触及并延伸了一些更为宏大的议题,包括乡村居住条件、物料资源、生态环境等。

In an era of media visualization, architects' practical achievements are usually disseminated and consumed in the form of images, while the motives, strategies, methods and social significance that revolve around the creation of architectural works are relatively easy to be ignored. To some extent, architects should maintain critical thinking.

In addition to the observation of the society, the environment, the city and the countryside, architects' design process still involves finding and raising questions from their perspectives by synthesizing knowledge far beyond professional scope, and according responding and discussing in a practical way. Only in this way can we produce radically innovative design based on the current situation.

Starting from basic questions of architecture, Atelier Heimat shows four micro-projects(Heyang Half-house, Post Road Corridor Bridge renovation, Luotuo Bay Bridge renovation, Cave of Sound) in this exhibition further focus on some major issues, such as rural living conditions, material resources, ecological environment, etc.

骆驼湾桥改造

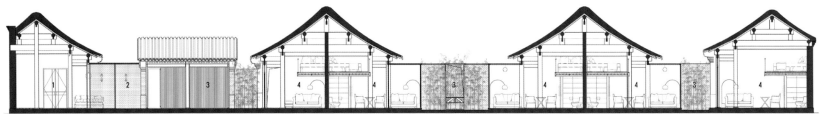

1. 接待
2. 洗衣房
3. 庭院
4. 客房

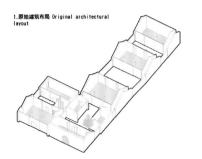

1. 原始建筑布局 Original architectural layout

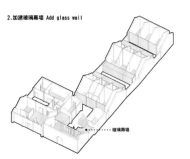

2. 加建玻璃幕墙 Add glass wall

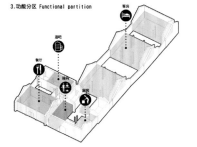

3. 功能分区 Functional partition

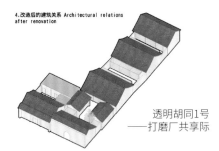

4. 改造后的建筑关系 Architectural relations after renovation

透明胡同1号
——打磨厂共享际

申江海
Shen Jianghai

大观建筑设计
Daga Architects

透明胡同系列

胡同改造，一直是大观建筑研究与实践的课题。在与胡同居民的交流和生活过程中，建筑师能够感同身受地了解他们对于生活空间的需求，能真切地感受到他们对于胡同文化的热爱、对于历史记忆的珍惜。

近两年，大观建筑在对胡同的旧房改造上做了很多尝试，以期在传统与现代之间找到一种平衡，在提升业主生活品质的同时，使他们在局促的小空间中获得更丰富独特的空间体验。胡同改造的意义并不仅在当下，更多的是着眼于未来，对新老结合方式的可能性进行探索。

"玻璃盒子"的概念被植入老建筑内，通透的玻璃幕墙能保证住宅全天候的充足采光，这不仅大大提升了小空间的采光率，还能够最大限度地削弱小空间带来的压抑感。同时，大面积的落地窗能够使古城胡同建筑的景观价值最大化。玻璃盒子的引入，在打破封闭隔阂的同时，给不同属性的空间带来更多的可能性。从材质属性上来看，光滑剔透的玻璃盒子与古老厚重的青砖灰瓦形成强烈的新老对比，从而隐喻了历史与现代的冲突与共生。

玻璃自身的透明特性不会遮挡原有建筑物，也不会减损它的美观。相反，它不仅符合使用者对阳光的追求，更能够从视觉上以及空间原理上使原有传统建筑的地位不被颠覆。设计采用了玻璃盒子的形式，再加上光线和人的运动，使整个空间充满活力。阳光穿过通透的玻璃倾泻到住宅内，给室内带来了温暖和灵动之感。

目前，大观建筑已经在北京的胡同内实践了三座透明胡同：透明胡同1号——打磨厂共享际；透明胡同2号——大小九之家；透明胡同3号——东四胡同。这三个项目都在持续探索现代与传统建筑的结合性问题，以期创造更多空间上的可能性。

Transparent Hutong Series

Hutong renovation has always been the study and practice subject of DAGA Architects. By the communication and living with residents in hutongs, architects can truly understand their needs for living space, can also feel their love for hutong culture and cherish historical memory.

In the past two years, DAGA Architects has made many attempts in the renovation of old houses in hutongs, aiming at finding a balance between contemporaneity and tradition. While the life quality of the owners is improved, they can also get more abundant and unique experience in the cramped small space. The significance of hutong renovation is not only in the present, but also in the future.. It focuses more on exploration of the possibilities of integration on classic and new styles.

The concept of "glass box" is embedded in the old buildings, the transparent glass curtain walls ensure adequate lighting throughout the day, and therefore lighting efficiency of small space is greatly improved. At the same time, large French windows can maximize the landscape value of the ancient hutong buildings. The introduction of glass box breaks the barrier and brings more possibilities to spaces with different attributes.. Seen from the attributes of materials, the smooth and translucent box forms a strong contrast with the old black bricks and gray tiles, thus metaphorizing the conflict and symbiosis between the history and modern era.

The transparency of glass itself will not block the original building or detract from its beauty; instead, it not only conforms to the user's pursuit for sunshine, but also safeguards the status of the original historical building against possible visual or spatial damage. The design takes the form of a glass box, coupled with the play of light and human movement, endowing the whole space with vitality. Pouring down through the lucent glass upon the building, the sun brings warmth as well as a sense of vividness to the interior space.

At present, DAGA Architects has practiced three transparent hutong works in Beijing's hutong area: Transparent Hutong No.1—West Grinding Factory; Transparent Hutong No.2—Twins House; Transparent Hutong No.3—Dongsi Hutong. These three projects continue to explore the combination of modern and traditional buildings and create more spatial possibilities.

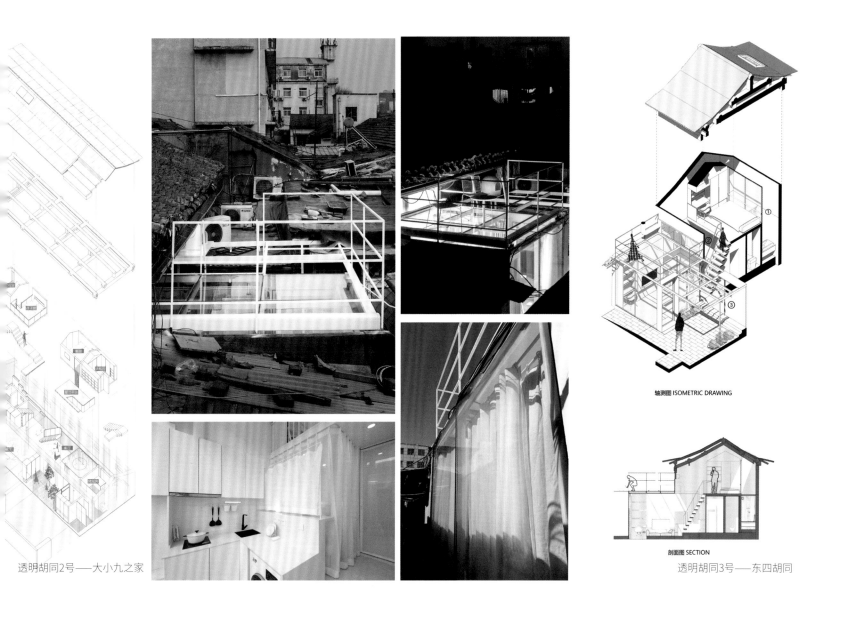

透明胡同2号——大小九之家

轴测图 ISOMETRIC DRAWING

剖面图 SECTION

透明胡同3号——东四胡同

后记
Postscript

本书于2019年12月初开始撰写。当时天津国际设计周建筑展（简称设计周建筑展或建筑展）总策展人宋昆教授携团队主要成员与天津国际设计周执行主席李云飞先生在巷肆创意产业园问津书院进行了一次深入的交流，商讨第四届天津国际设计周建筑展的具体策划方案。2020年5月初举行的设计周建筑展将邀请来自德国、意大利、日本等国家以及来自国内的杰出青年建筑师，他们以线上线下相结合的方式共同亮相该次展览，并以建筑师各自事务所为平台进行全球直播。策展团队希望以这种方式不断加强中外青年建筑师之间的交流与互动，进一步提升建筑展的国际化程度。最后，李云飞先生建议对2017—2019年设计周建筑展的成果进行梳理和总结，并结集出版，用图书留存珍贵的记忆。同时，本书也可以成为天津申请联合国教科文组织全球创意城市网络——设计之都的重要支撑材料。令人遗憾的是，就在策展方案逐步实施的过程中，新型冠状病毒肺炎疫情暴发，席卷全球并愈演愈烈。出于公共卫生安全方面的考虑，天津国际设计周自2020年起至今暂时在线上举办，建筑展也只好暂停，本书便成了设计周建筑展的阶段性总结。

2014年5月，天津国际设计周在天津市北宁公园首次举办，其中与建筑设计相关的板块包括天津国际设计周竞赛和国际设计大师班。2017年，天津国际设计周建筑展作为以从业建筑师和建筑专业教师为创作群体的独立主题展首次亮相设计周，并成为其重要的板块之一。建筑展总策展人为天津大学建筑学院宋昆教授，三年来参与策展工作的团队成员包括张昕楠、胡子楠、赵伟、赵劲松、苑思楠、冯琳、黄元炤、李德新等。首届建筑展的主题为"当代中的非当代"，展出的作品是建筑师创作的实体装置。2018年、2019年的主题分别为"此间""汇流"，展览内容为国内青年建筑师的近期设计作品。纵观天津国际设计周建筑展三年来的发展历程，其主题非常鲜明，始终聚焦我国城乡发展过程中的前沿问题，参展人员以京津冀地区崭露头角的中青年建筑师为主，随着参展建筑师人数不断增加，展览规模不断扩大，展览正逐渐成为该地区具有一定影响力的建筑专题展览。

本书的撰写工作历时两年完成，三位作者共同议定了本书的写作思路与整体框架，宋昆和胡子楠负责建筑展的写作与编辑，李云飞从天津国际设计周的视角予以指导，并慷慨资助了本书的出版。

在本书即将问世之际，作者团队首先感谢所有的参展建筑师，他们是鲍威、卞洪斌、卜骁骏、陈天泽、狄韶华、顾志宏、关英健、郭海鞍、韩文强、胡子楠、刘文娟、那日斯、任军、申江海、宋昆、田恬、屠雪临、王宽、王淼、王求安、王振飞、徐强、张大昕、张东光、张华、张继元、张曙辉、赵劲松、甄明扬、庄子玉、卓强（按姓氏拼音首字母排序），他们的专业水准与辛苦付出正是历年建筑展得以高品质呈现的重要基础，同时他们也为这本著作提供了基础性的图纸与资料。特别感谢意大利设计师李保罗先生，他作为本书的平面设计师，在图片甄选、书籍排版、装帧设计等方面做了大量工作，特别是因为疫情，他无法飞抵中国工作，只能在意大利通过网络连线的方式与作者团队沟通解决书籍版式设计中细致又繁复的诸多问题，难度可想而知。感谢ADA（建筑设计艺术）研究中心黄元炤先生为本书2019年建筑展部分的综述提供的基础性资料。感谢天津国际设计周组委会的工作人员，包括刘梦薇女士，她负责天津设计周组织团队与建筑展参展团队的对接，在本书写作的两年中，她做了大量的联系与协调工作；徐雅楠女士，她擅长意大利语，负责作者团队与意大利设计师李保罗先生的联络与沟通工作；孙一琨女士，她负责2017年建筑展部分参展作品的英文翻译工作。感谢天津理工大学语言文化学院孙晓晖老师在英文翻译工作中提供的帮助；感谢天津大学出版社韩振平、刘博超、朱玉红等编辑，正是他们的耐心、细心与责任心，才使得本书顺利出版。天津大学建筑学院博士生景琬琪、李康也参与了资料收集、图片整理等工作，在此一并表示感谢。

本书承蒙世界著名建筑与工业设计大师、天津国际设计周总顾问黑川雅之先生慨允作序，黑川先生从博大的文化视角对本书的出版予以肯定；感谢意大利那不勒斯费德里克二世大学前任校长马西莫·马雷利教授的序言，他从"共享文化"的概念谈到本书的读后感，为本书的后续思考提供了新的方向。

本书对天津国际设计周建筑展以及地区建筑文化现象做了一些基础研究，属于阶段性成果。书中仍有许多不足之处，有待专业学者与各位读者批评指正。未来，天津国际设计周建筑展将继续聚焦我国城乡建设领域的热点与难点问题，以前瞻性的视角引领建筑设计创新思维的导向，并努力成为展示我国建筑师群体创作现状的重要国际化平台。我们也期待在后疫情时代，天津国际设计周以更加新颖多元的方式出现在公共视野中，融入城市发展，走进百姓日常生活，为天津城市文化的构建、公众审美素养的提升以及地区建筑特色的塑造做出应有的贡献。希望通过我们的努力与坚持，天津国际设计周能够逐步成为这座城市的一个传统与品牌，愿我们的城市更美好！

宋昆 李云飞 胡子楠
写于天津问津书院
2022年5月

The writing of this book originated in early December 2019. At that time, Professor Song Kun, Chief Curator of Architecture Exhibition for Tianjin International Design Week (the Architecture Exhibition of the Design Week or the Architecture Exhibition for short), together with key members of his team, had an in-depth communication with Mr. Li Yunfei, the Executive Chairman of Tianjin International Design Week, at Wenjin Academy of Tianjin Xiangsi Creative Industrial Park to discuss the specific planning for the Architecture Exhibition of the 4th Tianjin International Design Week. The Architecture Exhibition of the Design Week held in early May 2020 will invite outstanding young architects from Germany, Italy, Japan and other countries, as well as from China, to present the exhibition together by an online-and-offline way, with respective architects' offices as the platforms for global live broadcast online. The curatorial team hoped to continuously strengthen the communication and interaction between young architects from China and abroad in this way, and further enhance the degree of internationalization of the exhibition. Finally, Mr. Li Yunfei suggested that the achievements of the 2017-2019 Architecture Exhibition should be sorted and summarized, collected and published, so as to use the book to retain the precious memory. At the same time, the book could also become an important support material for Tianjin to apply for the City of Design, which is the global creative city network of United Nations educational scientific and cultural organization. Regrettably, while the curatorial plan was being implemented, COVID-19 broke out, sweeping the world and intensifying. For public health and safety reasons, Tianjin International Design Week has been temporarily held online since 2020, and the Architecture Exhibition has to be suspended. This book really becomes a stage summary of the Architecture Exhibition of the Design Week.

In May 2014, Tianjin International Design Week was first held in Beining Park, and the sections related to architectural design included Tianjin International Design Week Competition and the International Design Master Class. In 2017, Architecture Exhibition of Tianjin International Design Week made its first appearance in the Design Week as an independent theme exhibition featuring professional architects and architectural teachers as creative groups, and became one of its important sections. The Chief Curator of the Architecture Exhibition is Professor Song Kun from the School of Architecture, Tianjin University, the team members who have been involved in the curatorial work for three years include Zhang Xinnan, Hu Zinan, Zhao Wei, Zhao Jinsong, Yuan Sinan, Feng Lin, Huang Yuanzhao and Li Dexin. The theme of the first exhibition is "Non-Contemporary in Contemporary", and the works on display are physical installations created by architects. The themes of the exhibitions in 2018 and 2019 are respectively "In Between" and "Confluence", and their content is about the recent design works of domestic young architects. Throughout the development of the Architecture Exhibition for the three years, its themes are very distinctive, and it always focuses on the cutting-edge issues in the process of urban and rural development in China. The exhibitors are mainly young and middle-aged architects emerging from the Beijing-Tianjin-Hebei region, and the number of exhibiting architects is increasing, and the scale of the exhibition is expanding, which makes it gradually becoming an influential architectural exhibition in the region.

The writing of this book took two years, and three authors jointly discussed on the idea and overall framework of this book. Song Kun and Hu Zinan were responsible for writing and editing. Li Yunfei provided guidance from the perspective of Tianjin International Design Week and generously funded the publication of the book.

As the book is about to be published, the author team would like to thank all the exhibiting architects firstly, who are Bao Wei, Bian Hongbin, Bu Xiaojun, Chen Tianze, Di Shaohua, Gu Zihong, Guan Yingjian, Guo Haian, Han Wenqiang, Hu Zinan, Liu Wenjuan, Na Risi, Ren Jun, Shen Jianghai, Song Kun, Tian Tian, Tu Xuelin, Wang Kuan, Wang Miao, Wang Qiu'an, Wang Zhenfei, Xu Qiang, Zhang Daxin, Zhang Dongguang, Zhang Hua, Zhang Jiyuan, Zhang Shuhui, Zhao Jinsong, Zhen Mingyang, Zhuang Ziyu and Zhuo Qiang (sorted by alphabetical order by last name initials), their professional level and hard work are the basis for the high quality of the architectural exhibitions, and they also provided the basic drawings and materials for this book. Special thanks to the Italian designer Mr. Paolo Altieri, as the graphic designer of the book, he has done a lot of work in pictures selection, format layout, graphic design and so on. Because of COVID-19, he was unable to fly to China for work, and had to communicate with the author team about so many detailed and complicated problems of the book design by internet in Italy, which was very difficult. Thanks to Mr. Huang Yuanzhao form ADA (Architectural Design Art) Research Center for providing the basic information on the overview of 2019 Architecture Exhibition. Thanks to the staff of the Tianjin International Design Week Organizing Committee, including Ms. Liu Mengwei, who was responsible for the relation between the Design Week organizing team and the Architecture Exhibition curator team, and has done a lot of contact and coordination work during the two years of writing this book; Ms. Xu Yanan, who is good at Italian, and was responsible for the contact between the author team and Italian designer Mr. Paolo Altieri; Ms. Sun Yikun, who was responsible for the English translation of some works introduction of 2017 Architecture Exhibition. Thanks to Ms. Sun Xiaohui from School of Languages and Culture, Tianjin University of Technology, for her help in English translation work. Thanks to Mr. Han Zhenping, Liu Bochao, Zhu Yuhong and other editors form Tianjin University Press, because of their patience, care and responsibility, the book can be published successfully. Thanks to Jing Wanqi and Li Kang, who are the doctoral students of the School of Architecture, Tianjin University, and participated in the collection of information and pictures.

Thanks to Masayuki Kurokawa, the world-famous architectural and industrial design master and the general consultant of Tianjin International Design Week, he generously wrote the book's preface and admired the publication of this book from a broad cultural perspective. Thanks to Professor Massimo Marrelli, who was the former rector for University of Naples Fedrico II, and he talked about the book from the concept of "shared culture" and provided a new direction for the subsequent thinking of the book.

This book has done some basic research on Architectural Exhibition of Tianjin International Design Week and the regional architectural culture phenomenon, which is a stage achievement. There are still many shortcomings in the book, which need to be criticized and corrected by professional scholars and general readers. In the future, Architecture Exhibition of Tianjin International Design Week will continue to focus on hot and difficult issues in the field of urban-rural development in China, leading the new orientation of innovative thinking in architectural design with a forward-looking perspective, and striving to become an important international platform to showcase the current state of creation of Chinese architects. In the post-COVID-19 era, we expect that Tianjin International Design Week will be presented to the public by more innovative and diversified way, and integrated into the development of the city and the daily life of the people, making due contributions to the construction of Tianjin's urban culture, the improvement of public aesthetic attainment and the creation of regional architectural characteristics. We hope that Tianjin International Design Week will gradually become a tradition and a brand of this city by our efforts and persistence. Let's make our city become more beautiful!

Song Kun, Li Yunfei, Hu Zinan

Writing at Tianjin Wenjin Academy

2022/05

策划编辑　韩振平工作室
组稿编辑　韩振平　朱玉红
责任编辑　刘博超
装帧设计　[意]李保罗（Paolo Altieri / Altieri Associati）

图书在版编目(CIP)数据

天津国际设计周建筑展：2017—2019 / 宋昆, 李云飞, 胡子楠著. -- 天津：天津大学出版社, 2022.5
 ISBN 978-7-5618-7100-3

Ⅰ.①天… Ⅱ.①宋… ②李… ③胡… Ⅲ.①建筑设计－作品集－世界－2017-2019 Ⅳ.①TU206

中国版本图书馆CIP数据核字(2021)第273168号

书　　名	天津国际设计周建筑展（2017—2019） TIANJIN GUOJI SHEJIZHOU JIANZHUZHAN（2017-2019）
出版发行	天津大学出版社
地　　址	天津市卫津路92号天津大学内（邮编:300072）
电　　话	发行部:022-27403647
网　　址	www.tjupress.com.cn
印　　刷	北京盛通印刷股份有限公司
经　　销	全国各地新华书店
开　　本	787mm×1092mm　1/12
印　　张	15 2/3
字　　数	400千
版　　次	2022年5月第1版
印　　次	2022年5月第1次
定　　价	159.00元

凡购本书，如有缺页、倒页、脱页等质量问题，烦请与我社发行部门联系调换

版权所有　　侵权必究